Comparison of Alternative Methods for Recharge of a Deep Aquifer

The mission of the Awwa Research Foundation (AwwaRF) is to advance the science of water to improve the quality of life. Funded primarily through annual subscription payments from over 1,000 utilities, consulting firms, and manufacturers in North America and abroad, AwwaRF sponsors research on all aspects of drinking water, including supply and resources, treatment, monitoring and analysis, distribution, management, and health effects.

From its headquarters in Denver, Colorado, the AwwaRF staff directs and supports the efforts of over 700 volunteers, who are the heart of the research program. These volunteers, serving on various boards and committees, use their expertise to select and monitor research studies to benefit the entire drinking water community.

Research findings are disseminated through a number of technology transfer activities, including research reports, conferences, videotape summaries, and periodicals.

Comparison of Alternative Methods for Recharge of a Deep Aquifer

Prepared by:
William F. Hahn and **Heather Thompson**
Boyle Engineering Corporation, Lakewood, CO 80228
and

Jeff Forbes and **Mark Ankeny**
Daniel B. Stephens & Associates, Inc., Albuquerque, NM 87109

Jointly sponsored by:
Awwa Research Foundation
6666 West Quincy Avenue, Denver, CO 80235-3098
and

El Paso Water Utilities
El Paso, TX 79961-0001
and

U.S. Department of the Interior, Bureau of Reclamation,
El Paso Field Division, El Paso, TX 79901

Published by:

Awwa Research Foundation
Advancing the Science of Water®

IWA Publishing

DISCLAIMER

This study was funded by the Awwa Research Foundation (AwwaRF). AwwaRF assumes no responsibility for the content of the research study reported in this publication or for the opinions or statements of fact expressed in the report. The mention of trade names for commercial products does not represent or imply the approval or endorsement of AwwaRF. This report is presented solely for informational purposes.

Published by IWA Publishing, Alliance House, 12 Caxton Street, London SW1H 0QS, UK

Telephone: +44 (0) 20 7654 5500; Fax: +44 (0) 20 7654 5555; Email: publications@iwap.co.uk;
Web: **www.iwapublishing.com**

AwwaRF report number 90962F. Originally published by AwwaRF for its subscribers in 2003.
IWA Publishing version published 2004.

ISBN 1 84339 868 0

TABLE OF CONTENTS

LIST OF TABLES

LIST OF FIGURES

FOREWORD

The Awwa Research Foundation is a nonprofit corporation that is dedicated to the implementation of a research effort to help utilities respond to regulatory requirements and traditional high-priority concerns of the industry. The research agenda is developed through a process of consultation with subscribers and drinking water professionals. Under the umbrella of a Strategic Research Plan, the Research Advisory Council prioritizes the suggested projects based upon current and future needs, applicability, and past work; the recommendations are forwarded to the Board of Trustees for final selection. The foundation also sponsors research projects through the unsolicited proposal process; the Collaborative Research, Research Applications, and Tailored Collaboration programs; and various joint research efforts with organizations such as the U.S. Environmental Protection Agency, the U.S. Bureau of Reclamation, and the Association of California Water Agencies.

This publication is a result of one of these sponsored studies, and it is hoped that its findings will be applied in communities throughout the world. The following report serves not only as a means of communicating the results of the water industry's centralized research program but also as a tool to enlist the further support of the nonmember utilities and individuals.

Projects are managed closely from their inception to the final report by the foundation's staff and large cadre of volunteers who willingly contribute their time and expertise. The foundation serves a planning and management function and awards contracts to other institutions such as water utilities, universities, and engineering firms. The funding for this research effort comes primarily from the Subscription Program, through which water utilities subscribe to the research program and make an annual payment proportionate to the volume of water they deliver and consultants and manufacturers subscribe based on their annual billings. The program offers a cost-effective and fair method for funding research in the public interest.

A broad spectrum of water supply issues is addressed by the foundation's research agenda: resources, treatment and operations, distribution and storage, water quality and analysis, toxicology, economics, and management. The ultimate purpose of the coordinated effort is to assist water suppliers to provide the highest possible quality of water economically and reliably. The true benefits are realized when the results are implemented at the utility level. The foundation's trustees are pleased to offer this publication as a contribution toward that end.

The research project described in this report was among the first to be funded under the foundation's Tailored Collaboration program. The Tailored Collaboration program was developed in 1997 to provide for subscriber involvement in regional research projects or projects that address an issue of interest to a significant subgroup of AwwaRF subscribers. The program is also designed to provide drinking water professionals access to and benefit from research results that might otherwise go unpublished. This Tailored Collaboration study, done in conjunction with EPWU, investigated the effectiveness of alternate techniques to recharge a deep, unconfined aquifer. Groundwater recharge and replenishment has received increasing attention from water utilities and other users in areas with limited supply, particularly in the arid southwestern U.S., as continued population growth and agricultural use have overdrawn and threatened the long-term sustainability of groundwater resources. The study addresses technical challenges to aquifer recharge in a hydrogeologic setting common to the southwestern U.S. and exemplifies the regional focus and emphasis of the Tailored Collaboration program.

Edmund G. Archuleta, P.E.
Chair, Board of Trustees
Awwa Research Foundation

James F. Manwaring, P.E.
Executive Director
Awwa Research Foundation

ACKNOWLEDGMENTS

The authors of this report are indebted to EPWU for their funding, cooperation and participation in this project. In addition, the advice and guidance of Mike Landis, U. S. Bureau of Reclamation, El Paso, Texas and the Project Advisory Committee (PAC), including Charles Heywood, United States Geological Survey, Richmond, Va.; Doug Munch, St. Johns River Water Management District, Palatka, Fla.; and Floyd Marsh, Scottsdale, Ariz. is appreciated.

EXECUTIVE SUMMARY

Water utilities, particularly those in arid western states, are being forced to seek increasingly more creative solutions for storing and reusing water as a way to lessen demand for potable supply and to balance differences in timing of raw water supply and potable water demand. Aquifer storage and recovery is often a viable and cost-effective alternative to the more traditional solution of constructing surface water storage reservoirs.

Recharge through injection wells is a relatively straightforward and proven technology in northeast El Paso, Texas. However, injection wells are relatively expensive to construct and they require frequent redevelopment and maintenance to sustain their injection capacity. The recharge water supply is highly corrosive due to chlorination, in which case, rapid deterioration of the well casing and screen has proven to be a costly drawback. Surface recharge through basins and vadose zone wells has not been tested in this environment. This investigation assessed the feasibility of two methods of surface recharge, infiltration basins and vadose zone wells. Infiltration basins are ponds constructed to maximize the infiltration of water into the underlying soil. Vadose zone wells, also known as dry wells, are typically completed in unsaturated, coarse, gravelly sediments of the vadose zone to allow maximum infiltration rates. Dry wells do not penetrate the water table.

The area of investigation for this study is the Hueco Basin in northeast El Paso, Texas. The Hueco Basin is a primary source of water for the City of El Paso, Fort Bliss, Biggs Air Force Base, and private industries in the area, and the primary source of water for Ciudad Juarez, Mexico. Soils in the study area are typically gravelly soils with varying degrees of cementation and caliche. The area affords ample underground storage space and reasonably high assurances of long-term recovery of stored water. While these characteristics are conducive to surface recharge, there are two major factors that could potentially inhibit the performance of a surface recharge facility or dry well in this area. These factors are, 1) low permeability geologic deposits, including caliche and clay layers, that could impede the vertical migration of recharge water en route to the water table; and 2) the potential for degradation of recharge water quality due to the leaching of soluble minerals by the recharge water as it percolates through the vadose zone. This study focused on these issues and generated useful information on the relative merits of these alternative recharge methods with respect to construction, operations and maintenance, cost, performance, and water quality impacts.

To accomplish the objectives of this investigation, the project was implemented in six tasks: (1) Site Selection, (2) Exploratory Test Pits and Infiltration Testing, (3) Exploratory Boreholes and Monitor Well Construction, (4) Design and Construction of Pilot Recharge Facilities, (5) Recharge Facility Operations, and (6) Monitoring Methods and Data Collection.

Task 1: Site Selection

Three candidate sites for further investigation and testing of pilot recharge facilities were identified and evaluated. The Research Team reviewed basic data on soils, water quality conditions, and hydrogeologic conditions at each site. A site located adjacent to an existing El Paso Water Utilities Production Well (Well 41) was selected for further investigation. This site is in close proximity to an existing water supply for use in recharge testing and hydrogeologic conditions at the site appeared favorable for recharge. In addition, there were no known occurrences of contaminants or brackish water plumes.

Task 2: Exploratory Test Pits and Infiltration Testing

Exploratory test pit excavation was conducted to assess the variability of near-surface soil and hydraulic properties on a local scale at the site. The results of the test pit excavation and percolation testing were used to determine preliminary infiltration rates and guide the selection of locations for the pilot-scale recharge facilities. The test pit excavation confirmed the existence of a low permeability calcium carbonate cemented caliche layer, which was encountered from approximately two to five feet below ground surface. Based on an evaluation of percolation test results, the ponded infiltration rate for a shallow recharge basin was estimated to be two to three feet per day.

Task 3: Exploratory Boreholes and Monitor Well Construction

Following the exploratory test pit excavation, a detailed investigation of subsurface geologic, hydrologic, and geochemical conditions was conducted by drilling boreholes. Dual-tube percussion drilling was used to advance borings to a maximum depth of 424 feet at the site. Soil samples were collected and logged at regular intervals from each borehole. Electrical conductivity tests were performed on soil samples to estimate the salt content of various horizons, and provide a measure of the potential for degradation of recharge water quality during percolation through the vadose zone. Based on the soil sampling and testing program, it was estimated that the total dissolved solids concentration could increase by up to 20 percent beneath the basin. However, any increase in total dissolved solids concentration at nearby wells was expected to be transient and dissipate with time.

Upon completion of exploratory borehole drilling, three monitor wells were installed which were designed to allow measurement of soil moisture content, collect soil pore-water samples at discrete depths beneath each of the facilities, and monitor the rate of recharge advancement. Each well contained multiple piezometers screened at discrete depths. The piezometers were installed to monitor the accumulation of recharge water on impermeable layers and allow collection of water samples from perched water zones. Two deep monitor wells were constructed to approximately 400 feet, one adjacent to the dry well and one adjacent to the infiltration basin. An additional monitor well was constructed to a depth of 100 feet adjacent to the infiltration basin, to monitor the progression of the recharge front at shallow depths through the vadose zone.

Task 4: Design and Construction of Pilot Recharge Facilities

A pilot infiltration basin and dry well were constructed, operated and monitored over a period of one year to assess their performance and impact on the groundwater environment. A 0.5-acre infiltration basin (150 feet square) was excavated to a depth ranging from about 8 to 10 feet. A caliche layer from approximately two to eight feet below ground was completely removed to maximize the infiltration rate and minimize the mobilization of soluble salts at the surface.

The dry well consists of a 20-inch diameter borehole drilled to a depth of 172 feet using a cable tool rig. A cable tool rig was selected because it allows for well construction with minimal addition of drilling fluids and is capable of drilling a relatively large borehole. Upon completion of the borehole, 12.5-inch outside diameter steel casing was installed. Within the 12.5-inch well casing, a 4-inch PVC recharge pipe was installed, which extends from the surface to a depth of 160 feet. A choke was installed in the recharge pipe to minimize the potential for entrainment of air in the injection pipe. Air logging of the receiving formation occurs when air, entrained in the injection fluid, accumulates in the formation in the vicinity of a well. The accumulation of significant amounts of air in the saturated zone can cause a reduction in permeability and the injection capacity of the well.

Both the infiltration basin and dry well were equipped with remote terminal units to control recharge operations and record flows and water levels from a flow meter and transducer, respectively, at each facility. The control system is configured so that flows and water levels are recorded and transmitted to a data logger. Water levels in the basin and dry well are maintained between high and low water level settings using a flow control valve at each facility. If the water level exceeds the high setting, flow to the facility is shut off until the water level recedes and reaches the low setting. When the water level reaches the low control setting the flow control valve fully opens.

Task 5: Recharge Facility Operations

Recharge operations at both facilities were started in mid-July 2001 and continued for a period of approximately one year through July 2002. The source of recharge water for both facilities is tertiary-treated wastewater from the Fred Hervey Water Reclamation Plant. In general, flows to the basin varied from 600 to 1,500 gallons per minute (gpm). For the entire data collection period, the basin recharged water at an average rate of 511 gpm (average includes no-flow periods) while water levels in the basin averaged about 0.5 feet. Cumulative flow into the basin was approximately 283 million gallons (869 acre-feet), for an average inflow rate of 0.74 million gallons per day (mgd) over the data collection period which lasted 384 days. Based on an average wetted footprint of approximately 0.4 acre, this corresponds to an average infiltration rate of about 1.8 mgd/acre or 6.5 feet/day of free water column. Field monitoring results confirm the rapid downward migration of infiltrating water.

Task 6: Monitoring Methods and Data Collection

Water level measurements taken at the deep monitor well adjacent to the basin show a 3-foot rise in water level 28 days following start-up, which appears to indicate the arrival of the wetting front at the water table. If the rise in water level is attributable to basin recharge, this suggests a vertical velocity for the wetting front of 13.8 feet/day. This rate is lower than the predicted seepage velocity estimated from water application rates (20 feet per day), which is expected because the predicted rate assumes vertical seepage only with no lateral spreading and no low permeability clay or caliche layers. The close agreement between measured and predicted seepage velocities suggests that clay beds in the vadose zone are not significantly retarding the downward migration of water toward the water table.

Subsurface temperature profiles at thermocouples installed in the basin floor also indicate rapid seepage velocities. Thermocouples were installed at multiple shallow depths beneath the basin floor to monitor diurnal temperature fluctuations. The time lag and rate of attenuation in amplitude with depth of the diurnal temperature waves can be used to estimate rates of water infiltration beneath different portions of the basin. During the first 10 days following start-up, the time lags for temperature maxima and minima between the various depth thermocouples suggest vertical seepage velocities ranging from 12 feet/day to 94.5 feet/day. The wide range is expected because of the varying sediment permeabilities across the basin, and the relatively short distances over which the seepage velocities were measured.

Neutron logging was performed prior to start-up to obtain background moisture profiles in the vadose zone. Neutron logs were completed several times following recharge start-up to monitor the progression of the wetting front through the vadose zone. The neutron moisture logs confirm the rapid downward movement of water. Neutron moisture logging showed the development of several wet zones at various depths between the basin floor and the water table. These zones likely indicate saturated conditions due to the buildup of perched water on clay layers.

Recharge operations at the dry well were considerably more complicated. The quantity of water recharged and the limited operation of the dry well were insufficient to allow definite conclusions related to infiltration rates and water quality impacts. There were several intermittent periods that the dry well was off-line due to equipment problems, maintenance, and lack of supply from the Fred Hervey Water Reclamation Plant. The only accurate, relatively continuous period of data collection for the dry well was from August through early November 2001. A total of 7.3 acre-feet of water flowed into the dry well, mostly during the period prior to November 2001. The sustained hourly average flow into the dry well was approximately 40 gpm at the beginning of recharge operations and declined to just under 20 gpm through early November 2001.

Summary and Conclusions

There were a number of difficulties encountered during construction and operation of the dry well that could not be resolved during the course of this investigation. Accordingly, the results may not be representative of dry wells in this environment. The poor performance of the dry well may be attributable to several factors including: 1) well plugging as a result of the drilling method and ineffective development following construction, 2) biofouling of the wellscreen or filter pack, and 3) air entrainment or the accumulation and/or trapping of air in the injection interval.

Background groundwater samples were collected from the deep monitor wells at both the basin and dry well. Additional water quality samples and field parameters were collected periodically through the remainder of the project. Background results indicate that the groundwater underlying the study area is of the sodium-chloride-bicarbonate type, with pH of approximately 7.6, and total dissolved solids concentrations in the range of 440 to 560 milligrams per liter (mg/L). Fred Hervey Water Reclamation Plant effluent has a slightly higher total dissolved solids concentration (620 mg/L) as compared to the groundwater. Major ion concentrations in groundwater samples collected from the monitor wells remained relatively constant over the course of the project. This may reflect the fact that dilution of the product water with the underlying groundwater occurred. Total dissolved solids concentrations of groundwater samples were similar to the Fred Hervey Water Reclamation Plant effluent suggesting that limited soluble salts are present in the vadose zone beneath the infiltration basin. Based on the water quality data collected for this project, neither recharge facility appeared to have a measurable impact on water quality.

In summary, construction, operations and maintenance of the basin were considerably less difficult than the dry well. The performance of the basin in terms of the quantity of water recharged far exceeded the performance of the dry well. The total capital and operations and maintenance cost of the basin was substantially less than the dry well ($70,000 versus $135,000). The total cost per unit of recharge (including the operations and maintenance cost) was $4,500,000/mgd for the dry well versus $61,000/mgd for the basin. These are not considered long-range costs for planning purposes as they are based on the one-year pilot study and do not include long-term operations and maintenance costs.

The feasibility of using injection wells for recharge and recovery of water is demonstrated by more than 15 years of operation of the injection well recharge facility at the Fred Hervey Water Reclamation Plant. When comparing surface recharge facilities with El Paso's on-going recharge operations using deep injection wells, the following observations can be made. There are several advantages to the use of injection wells: evaporation losses are minimal; land requirements for facilities are modest; and the risks of contamination of recharge water by shallow, soluble deposits found within the unsaturated zone are minimized. The City's existing injection wells were designed to inject up to one million gallons per day on a near-continuous basis. The wells have performed satisfactorily, although they require frequent redevelopment and maintenance to sustain their injection capacity. The primary disadvantage is that injection wells are expensive to construct and maintain. The water being injection is chlorinated in a final step of the treatment process. This leads to a highly corrosive water which has caused premature failure in several of the wells. The cost for construction of an injection well, including instrumentation, controls, security fencing, a permanent pump for periodic surging and development, and a pump house is about $375,000. The cost per unit of recharge capacity for a single injection well is therefore about $375,000/mgd of recharge capacity, based on a sustained recharge capacity of about 1 mgd.

Surface recharge facilities can be cost effective compared to an injection well if adequate land is available and relatively inexpensive. The sustained recharge capacity of the infiltration basin is roughly 1.15 million gallons per day based on the results of this study, however, the real long-term capacity is likely much greater. The basin may have been capable of maintaining a higher recharge rate, however, supply from the Fred Hervey Water Reclamation Plant was a limiting factor, in which case the maximum infiltration rate was not attained. The total capital cost of the infiltration basin was about $70,000. Because the basin was located on property owned by EPWU there were no costs associated with land purchase. The cost per unit of recharge capacity for the basin is $61,000/mgd of recharge capacity, based on a sustained recharge capacity of about 1.15 mgd. These costs could be substantially higher if the recharge water supply requires additional treatment. The source of water for this project was tertiary-treated wastewater, which required no additional treatment prior to recharge.

Maintenance for the basin typically consists of intermittent earthwork for reshaping and stabilizing basin walls, cleaning of the pond bottom and removal of low-permeability fine sediments. Maintenance costs for the basin are estimated to be about $5,000 per year. Maintenance of an injection well is performed on an as-needed basis, averaging about once every five years. Maintenance activities include surging, swabbing, acidization, and mechanical brushing to remove plugging and blockage of the well screens. Maintenance costs for an injection well are estimated to be about $30,000 every five years, or about $6,000 in terms of an annual cost. Accordingly, maintenance costs for these recharge facilities are substantially the same.

In summary, infiltration basins appear to be a viable alternative to use of injection wells for recharge of the deep aquifer. Based on data collected in this study, they appear to perform well, are relatively inexpensive to construct, have modest maintenance costs, and are easy to operate. Caliche layers and clay beds in the vadose zone are not significantly retarding the downward migration of water toward the water table. In addition, water quality sampling results to date have not indicated any significant deterioration of recharge water or native groundwater.

The operation of the dry well and the quantity of water recharged was generally insufficient to estimate sustained recharge and infiltration rates for comparison with the basin or injection wells. The dry well was shut down numerous times due to a lack of supply, equipment problems, or maintenance. The low infiltration capacity of the well may be attributable to some form of well fouling or ineffective development of the well following its construction. The performance of the dry well constructed for this project should not be considered indicative of the performance of dry wells in general.

CHAPTER 1
INTRODUCTION

PROJECT BACKGROUND

Water utilities, particularly those in arid western states, are being forced to seek increasingly more creative solutions for storing water as a way to balance differences in timing of raw water supply and potable water demand. Aquifer storage and recovery (ASR) is often a viable and cost-effective alternative to the more traditional solution of constructing surface water storage reservoirs for raw water storage. In the El Paso, Texas area, there are virtually no opportunities for constructing surface reservoirs, leaving aquifer storage as their only alternative. ASR has been considered by EPWU as a method to provide seasonal storage of surplus treated wastewater effluent and to help restore the Hueco Bolson (Basin) in northeast El Paso. The Hueco Basin is a primary source of water for the City of El Paso, Fort Bliss, Biggs Air Force Base, Ciudad Juarez, Mexico, and private industries in the area. Since 1903, groundwater levels have declined as much as 150 ft (Ashworth 1990). The area of investigation for this study is in the northern portion of a cone of depression that has developed around a major pumping center serving the City of El Paso. This area is located over an ancient water course of the Rio Grande River, and is well suited for both short- and long-term groundwater storage for several reasons. The area affords ample underground storage space. Long-term pumping in the Hueco Basin has lowered water levels and created a substantial depression in the water surface into which recharge water can be placed. Also, the area is underlain by a high permeability aquifer, which would support high capacity wells for both recharge and recovery.

The Hueco Basin has been successfully recharged at rates up to about 10,000 acre-feet per year (ac-ft/yr) through deep injection wells since the mid-1980s. The concept of subsurface storage of water as a means of sustaining and/or increasing the water supply available in the area northeast of El Paso was presented in the report: *Aquifer Storage and Recovery Investigation, September 1995* (Boyle Engineering Corporation and Parsons Engineering Science, Inc. 1995). That investigation concluded that the northeast El Paso area appeared to have conditions suitable for implementation of large-scale recharge. The area affords ample underground storage space and reasonably high assurances of long-term recovery of stored water. Long-term pumping in the

Hueco Basin has lowered water levels and created a substantial depression in the water table into which recharge water can be placed. The report also concluded that large-scale recharge provides an opportunity to mitigate aquifer overdraft and potentially restore groundwater supplies for continued use.

Two potential mechanisms for subsurface storage of water in the northeast El Paso area include well injection and surface recharge. The feasibility of using injection wells for storage and recovery of water is demonstrated by more than 15 years of operation of the recharge facility at the Fred Hervey Water Reclamation Plant (FHWRP). There are several advantages to storing water using injection wells, including essentially no loss of water due to evaporation and minimal land requirements for facilities. The principal disadvantage of injection wells is the significant cost associated with their construction. Conversely, surface recharge facilities can be cost effective if adequate land is available and relatively inexpensive, and where permeable sediments are present at or near the land surface.

While recharge through injection wells is a relatively straightforward and proven technology in this area, surface recharge through basins, infiltration galleries, and vadose zone wells (defined here to be wells which do not penetrate the water table, also known as dry wells) has not been tested. Two major factors that may impact the performance of a surface recharge facility are:

1. Low permeability geologic deposits, including caliche and clay layers, that could impede the vertical migration of recharge water en route to the water table; and
2. The potential for degradation of recharge water quality due to the leaching of soluble minerals by the recharge water as it percolates through the vadose zone.

Of particular concern are the cemented soils and caliche horizons that are often present near the surface in the El Paso region. In the vicinity of the study area, a caliche layer was encountered near the surface, which typically ranged from two to five feet thick. The feasibility of surface recharge depends in large measure on the hydraulic and chemical influences that the caliche layers may have on the recharge water. In the past, there have been several incidences in the northeast El Paso area in which groundwater has been degraded following extended ponding of water at the surface. While there is no direct information on the leachability of soils in the study area, recent investigations by the United States Geological Survey (USGS) taken in combination with drilling throughout the basin point to the existence of a highly soluble, laterally extensive caliche zone. Such a condition could render surface recharge facilities unfeasible.

This investigation initially identified three mechanisms for recharge, namely infiltration basins, infiltration galleries, and dry wells for evaluation and comparison. While an infiltration gallery requires less space than a basin and experiences no evaporation, it has a higher cost and is more difficult to maintain and operate than an infiltration basin. Based on concerns over the potential for water quality degradation, available funding, and EPWU staff resources, the Project Advisory Committee (PAC), and Research Team – consisting of Boyle Engineering Corporation and Daniel B. Stephens & Associates, Inc. – jointly agreed to eliminate the infiltration gallery from the study. A more detailed discussion of the elimination of the gallery from further investigation is provided in Appendix A. The methods of recharge selected for evaluation and testing were therefore limited to an infiltration basin and a dry well.

PROJECT PURPOSE AND NEED

The purpose of this project is to generate new and improved information on recharging a deep aquifer using infiltration basins and dry wells. Research completed during this project can be applied to similar environments where excessive pumping, limited recharge, and water quality problems have affected the aquifer system. Several potentially industry-wide benefits may be derived from this investigation, including:

- The exploratory drilling, sampling, and monitoring programs implemented during this project provide examples of methods useful in characterizing the unsaturated zone in areas being considered for recharge. Measurement and sampling of water in discrete vertical intervals can be applied to similar projects to define the progression of a recharge wetting front. Data collected during the sampling and monitoring program will be made available to EPWU to expand their existing knowledge and experience with surface recharge facilities.
- This project advances the understanding of geochemical changes to surface-recharged water in arid environments such as those in El Paso. The geochemical

data collected during this study will be particularly useful to EPWU in predicting water quality changes during passage of recharge water through the vadose zone. Investigation of subsurface water movement and water quality changes will be useful in assessing the feasibility of surface recharge projects in similar conditions.

- Methods used for data collection and monitoring recharge rates through the unsaturated zone can be used by other cities currently utilizing or considering groundwater recharge through surface recharge facilities.

RESEARCH OBJECTIVES

The objective of this research was to compare and assess the feasibility and applicability of two methods of recharge: infiltration basins and dry wells. The project generated useful information on the relative merits and deficiencies of these alternative recharge methods with respect to construction, operations and maintenance, cost, performance, and water quality impacts.

To accomplish the objectives of this investigation, the project was implemented in six tasks: (1) Site Selection, (2) Exploratory Test Pits and Infiltration Testing, (3) Exploratory Boreholes and Monitor Well Construction, (4) Design and Construction of Pilot Recharge Facilities, (5) Recharge Facility Operations, and (6) Monitoring Methods and Data Collection.

OVERVIEW OF RECHARGE METHODS INVESTIGATED

Two methods of recharge were selected for investigation and testing: an infiltration basin and dry well. An overview of these recharge methods is provided below.

Infiltration basins are ponds constructed to maximize the infiltration of water into the underlying soil. They differ in design and function from retention basins, which are intended to store and detain stormwater. Often, little infiltration occurs in retention basins because the basin floor tends to clog rapidly with fine sediments, creating a low-permeability layer that severely impedes downward flow of water. In contrast, infiltration basins are designed to minimize the development of a clogging layer. By allowing the basin to dry out completely between periods of infiltration and periodically grading and scarifying the basin floor and removing low-permeability fine sediments, high infiltration rates can be achieved. Intermittent cleaning of the basin floor is usually required. Graded fill material may be used in the pond bottom to create a filter for fine sediment. The primary advantages of infiltration basins include: (1) low initial cost and ease of construction, (2) large storage volume, and (3) possible beneficial effect on water quality (soil-aquifer treatment) depending on the chemical composition of the underlying soils. Disadvantages include: (1) significant land requirements, (2) health and safety risks or liability hazards, and (3) evaporative losses.

Dry wells are so-called because this type of well only extends into the vadose zone, above the water table. Dry wells have been used extensively in Arizona to control stormwater runoff. Dry wells are typically two to four feet in diameter, 20 to 50 feet deep, and are completed in unsaturated, coarse sediments to allow maximum infiltration rates. If the water source is expected to be turbid, a settling tank is often used at the surface to allow coarse sediment to settle out prior to injection into the well. While some degree of suspended and dissolved solids removal may be achieved during infiltration through the vadose zone, the transit time to the water table is usually relatively short. Therefore, as with any point source of recharge, it is important that the water infiltrating a dry well be of high quality to avoid contamination of the underlying aquifer.

Advantages of dry wells include large infiltration capacity and minimal land requirements. Disadvantages include high construction costs and the requirement that recharge water be of high quality to avoid well plugging and aquifer contamination.

PERMITTING AND REGULATORY ASPECTS OF RECHARGE METHODS

Requirements for permitting of recharge facilities vary depending on the type of recharge facility being contemplated (surface infiltration basin, infiltration gallery, or well), the quality of water used for recharge, and the state in which the facility is to be located.

Wells used for recharge and those operated for both recharge and recovery are regulated under the U.S. Environmental Protection Agency's (EPA) Underground Injection Control (UIC) Program. Such wells are considered Class V wells under this program in light of their function to "...replenish water in an aquifer." The primary regulatory responsibility for these wells lies with either the Federal UIC program, or with a state UIC regulatory agency where the state has assumed primary authority (UIC Primacy State). Where the UIC program is implemented directly at the federal level, states may retain additional jurisdiction through state water quality control agencies, state engineers, or regional water control or water management agencies. In Texas, aquifer recharge is regulated by the Texas Commission of Environmental Quality (formerly the Texas Natural Resource Conservation Commission). The Texas Commission of Environmental Quality's UIC program is guided by both state regulations/statutes and federal rules/statutes for the regulation of underground injection wells, and the protection of groundwater in Texas. In particular, recharge wells are regulated by Texas Administrative Code Title 30 Chapter 331 – Underground Injection Control.

Permitting of surface recharge facilities such as infiltration basins and galleries is typically regulated as a surface discharge source. Accordingly, responsibility for regulation and permitting of these facilities generally falls under the jurisdiction of state and/or local health and state and/or local water quality control agencies.

The direct use of treated effluent for recharge of underground sources of drinking water (USDW) is relatively uncommon. Accordingly, rules governing this type of activity are less uniform. In many cases, projects contemplating the use of treated effluent for recharge will be treated on a case-by-case basis. Direct injection of treated effluent into a USDW is prohibited in many states, even where such effluent has been treated to drinking water standards. In such cases, the use of surface facilities, such as infiltration basins, may be more easily permitted because of the greater separation between the point of recharge and the location of the USDW. In addition, depending on subsurface conditions, there is the potential for improvement in water quality as the recharge water moves through the unsaturated zone.

Permitting of the pilot recharge facilities installed under this project was accomplished by EPWU.

HYDROGEOLOGIC SETTING

The study area lies within the Texas portion of the Hueco Basin. It is bounded on the north by the New Mexico-Texas state line, to the west by the Franklin Mountains, to the south by the Texas-Mexico international border, and to the east by the eastward limit of fresh water within the Hueco Basin. The Hueco Basin extends from south-central New Mexico, through western

Texas, and into Mexico. The fresh water-bearing sediments consist primarily of interbedded layers of sand, gravel, clay, and silt which range from a few feet to several hundred feet in thickness near the center of the basin. The saturated thickness of the fresh water bearing zone varies and reaches a maximum thickness of up to about 900 ft near the structural center of the basin (Sperka 1996). The Hueco Basin is a primary source of water for the City of El Paso, Fort Bliss, Biggs Air Force Base, and private industries in the area, and the sole source of drinking water for Ciudad Juarez. Historically, some groundwater has been pumped for irrigation use in times of reduced surface water supply. Water levels in the area are 300 to 400 ft below ground surface. Since 1903, groundwater levels have declined as much as 150 ft.

Soils in the study area include three associations: the Del Norte-Canutio Association; the Turney-Berino Association; and the Hueco-Wink Association. Soils belonging to the Del Norte-Canutio Association occur in the foothills along the western margin of the basin, on nearly level to steep slopes. These are typically shallow, gravelly soils with varying degrees of cementation and caliche. Soils of the Turney-Berino Association occur on level terrain to gentle slopes. They contain a clay loam subsoil, are moderately deep, and contain caliche. Soils of the Hueco-Wink Association are found in level areas. They consist of fine sandy loam, are moderately deep in their formation, and typically overlie caliche.

Extensive drilling within the basin indicates that layers of caliche are pervasive. Recent investigations completed by the USGS in the study area, which include a deep exploration core hole, confirm this. While there has been no systematic mapping of the caliche, it has been encountered at various depths beginning at the surface and extending to the water table during well drilling conducted by EPWU.

CHAPTER 2
METHODS AND MATERIALS

PREVIOUS RESEARCH

Published papers and reports were consulted for information regarding prior and ongoing ASR investigations in the study area. These reports provided information on areas considered suitable for ASR, recharge method technology and design, and the relative success of ASR methods that have been implemented to date in this area and other similar environments. In addition, papers and reports providing information on groundwater resources and development and geologic and hydrologic data for the study area were reviewed. A bibliography of project resource materials is provided in Appendix B.

Three investigations of ASR within the study area were identified. The concept of subsurface storage of water as a means of sustaining and/or increasing the water supply available in the area northeast of El Paso was presented in the report: *Aquifer Storage and Recovery Investigation, September 1995* (Boyle Engineering Corporation and Parsons Engineering Science, Inc. 1995). That project resulted in the preliminary screening of four potential ASR sites in Las Cruces, New Mexico; West Mesa, New Mexico; the Canutillo Wellfield in Texas; and northeast El Paso in Texas. While ASR was considered technically feasible at all four sites, northeast El Paso was considered the preferred location.

After the completion of the 1995 study, supplemental investigations were conducted and presented in the report: *Aquifer Storage and Recovery: Supplemental Investigation of Recharge Basins 1997* (Boyle Engineering Corporation 1997). That study concluded that recharge capacity along the east margin of the Franklin Mountains in Texas was limited due to the presence of interbedded caliche layers, and water quality was a concern because of potential leaching of salts in the unsaturated zone as well as the possible mobilization of brackish water in the saturated zone.

The third investigation included a conceptual design of an ASR wellfield and collection facilities. The objective of that project was to recharge and recover 40 million gallons per day (mgd) of treated water from the Rio Grande River to meet a demand of 40 mgd in five service areas. The project contemplated about 70 ASR wells equipped for recharge and recovery.

Several papers were also reviewed that describe vadose zone well technology and pilot recharge testing. The concept of using vadose zone recharge wells for indirect injection of treated water in the City of Scottsdale, Arizona was addressed in the paper: Vadose Zone Recharge Wells: Field Evaluation of an Innovative Design (Marsh, Small, and Close 1997). Four vadose zone well designs were presented. Design differences included casing size, eductor system, and gravel pack material. The results of pilot tests were also presented, including an evaluation of multi-well operation, water quality changes, soil recharge capacity, and plugging potential. This paper provided useful information regarding the efficient operation of a vadose zone well based on the results of the pilot recharge testing. Other related papers include: Recharge Well Technology at the Water Campus Project, Scottsdale, Arizona (Marsh, Dueker, and Small 1993) and Vadose Zone Recharge Well Pilot Testing: An Innovative Option for ASR (Marsh, Small, and Close, undated). These papers were used as a reference for the design of the vadose zone well constructed for this project and the associated pilot recharge testing program.

SITE SELECTION (TASK 1)

The objective of this task was to identify a suitable site for further investigation and subsequent testing of pilot recharge facilities. The following three candidate sites were identified jointly with AwwaRF, EPWU, and the PAC.

1. Foothills site – This site lies along the eastern toe of the Franklin Mountains, approximately one mile south of the New Mexico-Texas state line, and west of Martin Luther King Highway.
2. Well 41 site – This site is located in close proximity to EPWU Production Well 41, lies approximately one mile east of Martin Luther King Highway, and is about three-fourths mile south of the Newman Generating Station.
3. Ponding area site – This site serves as a retention area for stormwater runoff and lies north of the intersection of Route 375 and Railroad Drive.

These sites are shown on Figure 2.1. Site selection criteria included the following:

- Proximity to water supply for use in recharge testing
- Proximity to existing water-supply wells and recharge wells
- Soils and surface geologic features
- Occurrence of known or suspected contaminant and/or brackish water plumes
- Prior history of recharge
- Presence of caliche or impediments to infiltration and percolation
- Projected depth to water
- Site access

To assist in the site selection process, the Research Team reviewed basic data on soils, water quality conditions, and hydrogeologic conditions for the three candidate sites identified. For each of the criteria presented above, a score ranging from 1 to 3 was assigned. A score of 3 is better than average or more favorable and a score of 1 is less than average or less favorable. Where data were insufficient for scoring purposes, a score of 2 was assigned. A total score was developed for each site, which is the sum of the scores for each criterion, as shown in Table 2.1. Each criterion was weighted equally.

Table 2.1

Scoring of candidate sites

Selection criteria	Foothills site	Well 41 site	Ponding area site
Availability of water supply	1	3	2
Proximity of site to existing wells	1	2	2
Favorable soils/geologic conditions	3	2	2
Proximity to known/suspected brine/contaminant plumes	1	2	2
Recharge history of site	2	2	3
Presence of caliche	2	2	2
Depth to water	2	2	2
Ease of access	2	2	1
Total score	**14**	**17**	**16**

As can be seen by the scoring presented in Table 2.1, the selection criteria produce relatively close total scores for the three sites. While there has been extensive drilling within the basin and along its western margin (foothills area), no site specific geologic or deep soils data could be found which would allow further discrimination of the sites, beyond conditions which may be inferred from their physical setting within the basin.

EPWU provided the Research Team with information that suggests the Foothills site overlies a brackish water plume of indeterminate dimensions. There was some risk that testing activities would mobilize this plume and subsequently impact nearby EPWU production wells. The Foothills site was eliminated from further consideration in favor of the alternate sites, given its close proximity to operational public supply wells.

On the basis of the total scores for each site, the Research Team selected the Well 41 site for further investigation.

EXPLORATORY TEST PITS AND INFILTRATION TESTING (TASK 2)

Exploratory test pit excavation was conducted at the Well 41 site to assess the variability of near-surface soil and hydraulic properties on a local scale. The results of the test pit excavation and percolation testing were used to evaluate the potential for surface recharge at the Well 41 site and help guide the selection of final locations for the pilot-scale recharge facilities.

Exploratory test pit excavation and infiltration testing was conducted on December 2 and 3, 1999 at the Well 41 site. The work involved the excavation of four test pits and percolation testing. The locations of the test pits (TP-1 through TP-4) are shown in Figure 2.2. EPWU provided equipment for the excavation of the test pits, which had dimensions of approximately 6 ft x 8 ft x 6 ft deep.

At each test pit, a detailed stratigraphic section was measured and sediment samples were collected at 1-foot intervals vertically for laboratory testing. Stratigraphic sections are shown in Figure 2.3. A similar stratigraphy was encountered in each of the test pits as follows:

- Layer 1: 0 to 2 ft deep: Silty sand, red-brown, dry, loose, more clayey towards base

- Layer 2: 2 to 5 ft deep: Caliche (calcisol), white, hard, gravel, sand and clay particles thoroughly cemented with calcium carbonate, becoming less cemented towards base
- Layer 3: 5 to 6+ ft deep: Fine to medium sand or silty sand, dry, loose to very weakly cemented, with minor fine gravel

At each test pit, a calcium carbonate cemented caliche layer was encountered from approximately two to five feet below ground surface (bgs). The caliche layer was removed to expose fine to medium sand or silty sand underlying the caliche. Although the top of Layer 3 was not encountered in TP-3, it is likely this unit is present at a slightly greater depth than was excavated. Layers 1 and 2 appeared to possess very low vertical hydraulic conductivities. In contrast, the sand encountered beneath the caliche (Layer 3) appeared quite permeable.

Falling-head percolation tests were performed to obtain a preliminary indication of potential infiltration rates at the site. The tests consisted of adding between 1,000 and 1,500 gallons of water to each of the four test pits (using a water truck), then monitoring the rate of water level decline as the water infiltrated into the sandy sediment exposed in the bottom of each pit. Percolation rates were monitored by periodically noting the water level (stage) on a wooden yardstick attached to a piece of steel rebar driven into the bottom of each test pit. The first set of percolation tests was performed on December 2, 1999. Each pit was filled to a depth of three feet at the start of the test. All four test pits were found to be dry the next morning, indicating that permeabilities were sufficiently high so that all of the water had infiltrated into the bottom of the test pits in less than 24 hours. A second set of infiltration tests was performed on December 3, 1999 using the same procedures. Because the first set of tests included an initial period of wetting the relatively dry sandy sediment (Layer 3), the results of the second set of infiltration tests are considered more representative of expected infiltration basin performance.

The results of the percolation tests are shown in Figure 2.4. If infiltration rates had remained constant during the tests, a linear decline in water level over time could be expected. Instead, the water level data fall along an exponential curve, indicating a small decrease in infiltration rate over time. This is to be expected because of the declining head (water level) and because infiltration rates should fall somewhat after the initial wetting period when the most rapid infiltration rates occur. Because the wetted sides of the test pits consisted primarily of low-permeability caliche (Layer 2), most of the infiltration of water is believed to have occurred as vertical flow into the underlying sand (Layer 3). Lateral flow into the test pit walls was considered to be minimal, though some lateral spreading of water in Layer 3 undoubtedly occurred. Based on an evaluation of the percolation test results, the ponded infiltration rates for a shallow recharge basin were estimated to be on the order of about two to three ft/day. The percolation test results for TP-4 indicate this rate could be significantly higher if the sand encountered below the caliche is a clean sand, as opposed to the silty sand encountered in the other test pits.

The falling head ponded infiltration data were used in an analytical solution to solve for the mean saturated hydraulic conductivity (K) over the wetted region beneath each test pit, following the method of Philip (1992). Figure 2.5 shows the results of this analysis. The early data collected were considered less reliable than the data generated toward the end of the tests. Therefore, the data from the end of the second day of infiltration tests were used to estimate the mean saturated hydraulic conductivity beneath each test pit as shown in Table 2.2.

Hydraulic conductivity values ranged from 4.2 ft/day (0.0010 cm/sec) to 18.0 ft/day (0.0064 cm/sec) for the sand at the base of each pit. As shown in Figure 2.5 and Table 2.2, the calculated hydraulic conductivity for TP-4 was significantly greater than that for the other test pits. This is consistent with the observation that the sand exposed in the bottom of TP-4 was clean sand and not the silty sand observed in the other test pits. While the short duration percolation tests performed did not reveal the presence of impeding layers that may be present at deeper depths, the results indicated favorable conditions for the construction of an infiltration basin.

Laboratory test results, including initial moisture content, dry bulk density, wet bulk density, calculated porosity, and particle size characteristics, were also conducted. The results of these tests are provided in Appendix C.

Table 2.2
Mean hydraulic conductivity

	Mean hydraulic conductivity	
Test pit	**(cm/sec)**	**(ft/day)**
TP-1	0.0015	4.2
TP-2	0.0017	4.8
TP-3	0.0010	2.8
TP-4	0.0064	18.0

Electrical conductivity (EC) tests were performed on soil samples collected from the test pits to estimate the quantity of soluble salts present (e.g. sodium chloride). This information can provide an indication of the potential for degradation of recharge water quality during percolation through the vadose zone. Soil EC tests were performed by mixing 50 milliliters of deionized water with 50 grams of dry soil to prepare a 1:1 soil slurry (mass basis). After allowing the slurry to equilibrate for 10 minutes, the water was passed through a Whatman No. 1 paper filter (nominal 11 micrometer pore size), and the EC of the filtrate measured using a Horiba ES-12 conductivity meter. The results of the EC tests are shown graphically in Figure 2.6. With the exception of TP-2, EC values for the 1:1 soil water extract were generally between 100 and 500 microsiemens per centimeter (μS/cm). Values for TP-2 ranged up to 2,935 μS/cm. EC test results indicate that the soluble salt content of soils near the surface varies considerably from one location to another. Much higher salt concentrations are present at shallow depths in the vicinity of TP-2. The highest salt concentrations at this location corresponded with samples collected near the base of the caliche unit.

Based on favorable results of the exploratory testing at TP-4, and proximity to a reclaimed water pipeline that runs north-south approximately 250 ft west of Well 41, preliminary locations were chosen for the dry well and infiltration basin, as shown in Figure 2.7. Proximity to the reclaimed water pipeline was necessary because it is the source of recharge water for the pilot facilities. The infiltration basin and dry well were located sufficiently far from Well 41 and each other to prevent significant interference between the facilities.

EXPLORATORY BOREHOLES AND MONITOR WELL CONSTRUCTION (TASK 3)

A detailed investigation of subsurface geologic, hydrologic, and geochemical conditions at the test site was accomplished by drilling exploratory boreholes. Upon completion of exploratory borehole drilling, monitor wells were installed to allow measurement of soil moisture content, soil pore-water quality at discrete depths beneath each of the facilities, and the rate of recharge advancement. Two deep monitor wells were constructed to approximately 400 ft, one adjacent to the dry well and one adjacent to the infiltration basin, as shown on Figure 2.7. These wells are referred to as Monitor Well 1 (M1) and Monitor Well 4 (M4). M1 and M4 were used to monitor vadose zone and saturated zone conditions at the dry well and infiltration basin, respectively. An additional shallow vadose zone monitor well was constructed to a depth of approximately 100 ft adjacent to the infiltration basin, and is referred to as Monitor Well 3 (M3), as shown on Figure 2.7. M3 was used to monitor the progression of the recharge front at shallow depths through the vadose zone. An additional shallow monitor well, labeled Monitor Well 2 (M2), was installed at the site proposed for the infiltration gallery; however, because the gallery was not constructed, monitoring activities were not conducted at M2.

Drilling Method

Several drilling methods were considered including sonic drilling, dual-tube percussion drilling, and hollow stem auger drilling. A method requiring minimal fluid addition was preferable to avoid altering in-situ moisture conditions and to minimize the need for well development. The AP-1000 dual-tube percussion rig was selected to advance the borings prior to installation of each monitor well because it offered a variety of advantages over other methods, including lower cost, faster drilling speeds, greater assurance of reaching the target depth (400 ft), and a larger diameter borehole. It also provides good stratigraphic definition and sample integrity during drilling because it does not require introduction of water or drilling mud.

Laboratory Test Results

An evaluation of the subsurface geologic, hydrologic, and geochemical conditions at the test site was accomplished through logging, sampling, and laboratory testing of subsurface materials. Soil grab samples were collected and logged at 2.5- to 5-foot intervals in all boreholes. In addition, split-spoon core samples were collected at approximately 10-foot intervals. This sampling frequency was judged adequate to define the stratigraphy and provided good samples for laboratory testing. Lithologic logs for monitor wells M1, M3, and M4 are shown in Figures 2.8, 2.9, and 2.10, respectively.

EC tests were performed on soil samples to estimate the salt content of various horizons, and provide a measure of the potential for degradation of recharge water quality during percolation through the vadose zone. A total of 72, 25, and 82 vadose zone sediment samples were tested from M1, M3, and M4, respectively. The EC values were determined on 1:1

soil:deionized water suspensions (by weight). The EC test results are included in Appendix D. Most of the high EC results correspond to either clay beds or caliche horizons. For example, in M1 a significant caliche layer was encountered between eight and fourteen feet below ground. EC test results for samples collected in that interval are approximately 3,500 μS/cm, which is about seven times the average EC for soil samples collected at M1. Similarly, a caliche layer was encountered in M3 between 41 and 44 feet below ground. EC test results for samples collected in that interval are approximately 250 μS/cm, which is almost two times the average EC for samples collected at M3. The EC test results for M4 are similar to M1 in that the higher EC measurements correspond to either clay beds or caliche zones. However, the peak EC test results for samples at shallow depths are significantly lower in M4 than M1.

Based on the soil sampling and testing program, the Research Team prepared a preliminary estimate of the potential for salt mobilization during the recharge testing period. The purpose of this evaluation was to provide a more quantitative estimate of the potential increase in salt concentration that might be observed at Well 41 from a pilot recharge facility. A summary of this evaluation is provided in Appendix E. It was estimated that the infiltration basin could increase TDS levels at Well 41 by up to 20 percent. An increase in TDS concentration at Well 41 and other nearby wells, however, was expected to dissipate with time following the end of recharge activities.

Monitor Well Design and Construction

Monitor wells were designed to measure soil moisture content and soil pore-water quality at discrete depths beneath each of the facilities and the rate of recharge advancement. Well completion diagrams for M1, M3, and M4 are shown in Figures 2.8, 2.9, and 2.10, respectively.

Each well was equipped with a 2.0- or 2.5-inch PVC neutron probe access tube for moisture monitoring. Moisture content was monitored using neutron moisture logging, a standard method used to monitor changes in moisture content at different depths in unsaturated materials. Data from these logs can provide an indication of whether the geologic materials surrounding the boreholes are losing water, gaining water, or are in a steady-state condition.

In addition to the neutron probe access tube, up to three 1-inch diameter PVC piezometers were completed to impermeable layers, which correspond to the tops of potential perching horizons, such as clay beds or caliche layers. The piezometers were designed to monitor the accumulation of recharge water on impermeable layers and permit collection of water samples from perched water zones expected to develop during recharge. A transducer and data logger were installed in piezometer 3 in M4 (M4/P3) to provide continuous monitoring of water levels in that piezometer.

Water level measurements were taken at both M1 and M4 after well construction was completed. The depth to water in M1 was 393.4 ft below top of casing. The depth to water in M4 was 386.1 ft below top of casing. All of the 1-inch piezometers in M1, M3, and M4 were dry.

DESIGN AND CONSTRUCTION OF PILOT RECHARGE FACILITIES (TASK 4)

A pilot infiltration basin and dry well were constructed, operated, and monitored over a period of one year, to allow a reliable assessment of their performance and impact on the groundwater environment. Construction and instrumentation of the facilities was completed by EPWU, Layne-Christensen, and Alpha-Southwest. The design, construction, and instrumentation of each facility are described in the following sections.

Infiltration Basin Design and Construction

A one-half acre infiltration basin was constructed approximately 1,000 ft south of Well 41 and 100 ft east of the reclaimed water pipeline, as shown in Figure 2.7. The basin measured about 150 feet by 150 feet. A typical cross section and plan view drawing of the infiltration basin is shown on Figure 2.11.

The basin was excavated with a large bucket excavator to a depth sufficient to expose the sand underlying the caliche layer. The depth of the basin ranges from about 8 to 10 feet. A caliche layer from approximately two to as much as eight feet below ground was completely removed to maximize the infiltration rate and minimize the mobilization of soluble salts at the surface. The southern half of the basin was over-excavated compared to design specifications. Coarse sand from the center of the basin was used to slope back and bench the basin walls to improve stability. Excavated materials were stockpiled around the perimeter of the basin to create an embankment at the basin edge. A 6-foot high perimeter fence was installed around the basin to restrict access to the site.

A 4-inch diameter stilling well was installed along the south wall of the basin west of the discharge pipe into the basin, as shown on Figure 2.11. The stilling well houses a transducer for monitoring water levels in the basin. A staff gage was attached to the outside of the stilling well for manual water level measurements and as a check against water levels recorded by the transducer.

Thermocouples were installed in the basin floor for temperature profiling in order to capture the diurnal thermal pulses associated with recharge. Monitoring of diurnal and seasonal temperature fluctuations at shallow depths provided estimates of temporal and spatial variations in recharge rates and vertical seepage velocities. The installation of the thermocouples is described in more detail under the section titled "Monitoring Methods and Data Collection (Task 6)."

The source of recharge water is tertiary-treated wastewater from the FHWRP. EPWU currently operates 10 injection wells for recharging tertiary treated wastewater from the FHWRP into the Hueco Basin aquifer. A reclaimed water pipeline runs north-south approximately 100 ft west of the infiltration basin. The reclaimed water pipeline is a 16-inch diameter asbestos cement pipeline that is approximately five feet deep, with operating pressures ranging from 35 to 85 pounds per square inch (psi). A 4-inch PVC pipeline was extended from the reclaimed water pipeline to the infiltration basin. A manually operated gate valve was installed directly after the connection with the reclaimed water pipeline for additional flow control to the basin and emergency shut-off purposes. A pre-cast concrete vault was installed along the 4-inch pipeline approximately 20 ft south of the basin. The concrete vault houses a totalizing flow meter and flow control valve.

A cinder block instrumentation shed was constructed adjacent to the basin and is used to house data acquisition equipment. The shed was necessary to protect against vandalism and to facilitate maintenance, operation of the control system, and storage of batteries used to power the instrumentation and control system.

It was initially planned to operate the control system using AC power at Well 41. EPWU subsequently determined that the cost for running power to the test site was prohibitive, and that an alternate source of power needed to be considered. The Research Team and EPWU jointly agreed that power would be supplied from rechargeable batteries. Based on the power load associated with the control system, six 12-volt batteries were purchased and housed in the instrumentation shed. Three batteries were switched out and recharged every two weeks.

A remote terminal unit (RTU) was installed in the instrumentation shed to control recharge operations at the basin and to record water levels and flow data transmitted from the basin flow meter and transducer, respectively. This system was chosen primarily based on its low cost and ease of operation. The system is configured so that flows to the basin and water levels recorded by the transducer are transmitted to a data logger in the instrumentation shed. The water level is compared to high/low water level settings, which are preset for the flow control valve. The flow control valve is set to operate in either a fully open or fully closed position. Therefore, if the water level exceeds the high setting the valve closes completely and does not open until the water level recedes and is lower than the low setting. Initial design considered use of a variable flow control valve, which would adjust the degree of valve opening in response to water level to maintain a more constant water level in the basin. This form of flow control requires significantly more power for operation and was abandoned in favor of the current design once it was determined that AC power would not be available.

The RTU archives and stores all of the real-time data within the RTU archiving module. The RTU server enables real-time data to be uploaded from the RTU to any laptop computer with the appropriate software. This allows real-time data to be obtained and transferred to a spreadsheet, database, or word processor program. In addition, real-time data is transmitted via telemetry to EPWU's office for remote data acquisition.

Dry Well Design and Construction

The dry well consists of a large-diameter well completed above the water table in the unsaturated sediments of the vadose zone. It was desirable to use a drilling method that requires minimal fluid addition and disruption of the native formation materials because of the subsequent difficulties in developing the well. A bucket auger rig was initially selected because that method does not require drilling fluids and is capable of constructing large diameter boreholes. Unfortunately, a drilling contractor could not be found with the requisite qualifications, experience, and licensing needed to construct the well as designed with a bucket auger rig. As a result, alternatives to bucket auger drilling were evaluated. The Research Team and PAC jointly agreed that the vadose zone well would be constructed using a cable tool rig. The cable tool rig allows for well construction with minimal addition of drilling fluids and is capable of drilling a relatively large borehole, which was not possible with other more conventional drilling methods.

The dry well was constructed by Layne-Christensen. Drilling started on January 19, 2001 and was completed on February 5, 2001. The borehole was drilled by advancing 20-inch steel drive casing to a depth of 172 ft. Approximately 4,000 gallons of water (treated effluent) were used during drilling. The water was added to create a slurry of drilling cuttings, which were bailed from the casing as it was advanced.

Upon completion of the borehole, 12.5-inch outside diameter steel casing was installed to a depth of 172 ft, as shown on Figure 2.12. Blank casing was installed from 0 to 92 ft, and from 162 to 172 ft bgs. The injection interval consists of 70 ft of 0.050-inch mill-slotted well screen from 92 to 162 ft bgs. The slotting pattern consists of 2-inch long x 0.050-inch wide slots, 24 slots per row, two rows per foot. Steel centralizers were installed every 40 ft to center the casing within the borehole. Once the 12.5-inch casing was installed, one-eighth inch gravel pack was added to the annulus as the 20-inch drive casing was retracted. The uppermost 20 feet of the 20-inch drive casing was left in the ground for surface completion. The gravel pack extends from the bottom of the well to a depth of 20 ft bgs. The uppermost 20 feet of the borehole was sealed with cement grout.

The well was equipped with a 4-inch Schedule 40 PVC recharge pipe extending from the surface to a depth of 160 ft bgs. The bottom of the drop pipe was fitted with an end cap, which has a 2.5-inch diameter hole. A 2-inch Schedule 40 PVC pipe was installed inside the 12.5-inch casing to house a transducer. The 2-inch transducer pipe extends from the surface to a depth of 160 ft bgs. The transducer was set at 159 ft bgs (although subsequently repositioned as described later in this report) and is used to measure the water level within the well. Figure 2.12 presents an as-built drawing of the dry well.

EPWU contracted with Alpha-Southwest to install surface facilities at the dry well. A wellhead detail showing the surface completion of the dry well is shown on Figure 2.13. Surface completion consisted of connecting the 6-inch PVC pipeline from the reclaimed water pipeline to the 4-inch PVC recharge pipe in the well. A pressure gage and 1-inch air release valve were installed at the wellhead.

Air logging of the receiving formation occurs when air, entrained in the injection fluid, accumulates in the formation in the vicinity of a well. The accumulation of significant amounts of air in the saturated zone around the injection site can cause significant reductions in permeability and the injection capacity of the well. While the risk of air logging is considerably less in an unconfined aquifer, a down-hole, retrievable choke was installed by Alpha-Southwest to minimize these risks. This device minimizes the potential for entrainment of air in the injection pipe by maintaining a pipe-full condition in the injection pipe.

The choke was fabricated using nominal 3.5-inch diameter steel pipe, measuring 1.5 ft in overall length. The tubing is open on the upper end. The lower end is threaded, and fitted with an end cap in which a centered hole has been drilled, to a diameter slightly smaller than the inside diameter of the choke tubing. A set of three circular orifice plates, having drilled openings measuring 1.0-inch, 1.5-inch, and 2.0-inch in diameter were machined to fit flush over the end of the threaded tubing and are held in place with the threaded end cap. The assembly was fitted with a bail hook on its upper end. The choke is lowered into the well on small diameter steel cable. It was designed to rest on the bottom end of the injection piping, which was also fitted with an end cap, into which a hole was drilled to a diameter somewhat smaller than the inside diameter of the injection piping. The upper end of the cable is secured at the wellhead to allow easy retrieval and installation of an alternate orifice plate.

The need for multiple orifice sizes stems from the difficulty of estimating in advance the performance of the well, and in particular the water level buildup that can be anticipated during injection. The well's performance has a direct bearing on the hydraulics of the injection piping. Alpha-Southwest installed the smallest diameter choke (1-inch) in the 4-inch recharge pipe for initial operation of the dry well.

The source of recharge water for the dry well is tertiary-treated wastewater from the FHWRP. A 6-inch PVC pipeline was extended from the reclaimed water pipeline to the dry well. A manually operated gate valve was installed directly after the connection with the reclaimed water pipeline for additional flow control and emergency shut-off purposes. A pre-cast concrete vault was installed along the 6-inch pipeline approximately 20 ft west of the dry well. The concrete vault houses a pressure reducing valve (PRV), totalizing flow meter, and flow control valve.

A cinder block instrumentation shed was constructed west of the dry well, which is used to house data acquisition equipment. The instrumentation and control system installed at the dry well is identical to the infiltration basin. The system is configured so that flows to the dry well and water levels recorded by the down-hole transducer are transmitted to a data logger in the instrumentation shed. The water level is compared to high/low settings, which are preset for the flow control valve.

RECHARGE FACILITY OPERATIONS (TASK 5)

Infiltration Basin Operations

Recharge into the infiltration basin was started at 1:20 pm on July 10, 2001 at a rate of approximately 675 gallons per minute (gpm). The low and high levels for the basin were set at one foot and two feet above the transducer, respectively.

Flow data and water level measurements were recorded manually from July 10, 2001 through mid-day July 12, 2001. The discharge rate into the basin ranged from 600 to 900 gpm during this start-up period. Manual water level readings were collected in lieu of electronic data because water had not yet reached the transducer in the stilling well. The maximum water level recorded at the basin staff gage was 0.92 ft. At this level, approximately two-thirds of the basin floor was wetted and four of the thermocouple locations were submerged. Successful transmission of discharge and water level data to a computer at the FHWRP began on July 12, 2001. Until water in the basin began to submerge the transducer, the electronic record indicated a "0.00" water level in the basin. Initially, discharge and water level data were collected every 10 seconds. This data recording interval was reduced to every minute after about one month of operation.

Hourly averaged flow rate and stage versus time data are presented in Appendix F (see Figure F.1) for the entire data collection period. There were several periods throughout the year that the basin was off-line due to equipment problems, maintenance, and lack of supply from the FHWRP. These periods are described in more detail under the section titled "Operational Difficulties and Potential Improvements". Figure 2.14 presents the periods during which the basin was in operation.

In general, flows to the basin fluctuated from 600 to 1,500 gpm. Based on predictions of available effluent from the FHWRP, the control system was initially programmed such that the maximum recharge rate that could be recorded was 1,000 gpm. Therefore, prior to mid-December 2001 the maximum recorded flow to the basin was 1,000 gpm; however, actual recharge rates exceeded that limit. The control system was reprogrammed in mid-December and a new flow meter was installed to record flows in excess of 1,000 gpm. Prior to mid-December, totalized flow meter recordings were relied on for measurements of flow into the basin.

The low and high water levels for the basin were kept at one and two feet, respectively, for the duration of the data collection period. While water levels were relatively constant from day to day, there was a strong diurnal fluctuation in water level, with the highest stage occurring close to midnight. The diurnal fluctuations in water level may be a result of variations in flow rate over the course of a day. The flow into the basin is generally higher at night due to the reduced demand for reclaimed water. The regularity of the fluctuations suggests they may also be attributable to daily temperature oscillations, with the resulting change in the viscosity of the water. As the water in the basin warms up during the morning hours, the viscosity of the water declines, causing an increase in the basin infiltration rate, and concomitant decline in water level. The daily fluctuations in basin open water temperature are in the range of 3 to 5°C.

Water depths in the basin gradually increased to about 1.0 to 1.5 ft through the end of November 2001. In mid-December 2001, the water level began to fluctuate regularly between the low and high settings of 1.0 and 2.0 ft, respectively. The water level increased more rapidly because of a buildup of algae on the basin floor, which reduced the infiltration rate. At one point, the basin stage increased to almost four feet. Normally, the high setting would have signaled the flow control valve to close, shutting down flow to the basin; however, damage to the transducer cable interfered with the signal to the flow control valve.

To address the build-up of algae on the basin floor, recharge operations were curtailed from December 17, 2001 through January 14, 2002 for maintenance work. A trackhoe was used to scarify the basin floor and remove the layer of algae that had accumulated at the surface. During that time the transducer cable was also repaired. After recharge operations were resumed on January 14, 2002, the basin stage dropped considerably and remained relatively constant at about 0.7 ft through March 2002. The basin was operated almost continuously from January 14, 2002 through March 2002. However, during that period about two weeks of flow and water level data were not collected because of battery failures. In April, recharge operations were interrupted numerous times due to maintenance work on a pump at the FHWRP and battery failures. Recharge operations at the basin were essentially suspended from May 2002 through late July 2002 due to lack of available reclaimed water from the FHWRP.

Dry Well Operations

Recharge operations at the dry well were started on July 11, 2001. The instrumentation and control system at the dry well was not fully operational at that time; therefore, the flow control valve was operated manually and flow and water levels were recorded manually during recharge start-up. To minimize the potential for air-entrainment, the smallest diameter choke (1-inch) was installed in the 4-inch recharge pipe. The initial flow rate into the well was approximately 180 gpm. The pressure gage on the recharge pipe indicated a pressure of 47 psi. The maximum water level in the well was 94 ft above the transducer or 65 ft bgs. The flow rate was adjusted after about one hour to 110 gpm, to investigate well performance at a reduced rate of injection. The maximum water level after reducing the flow was 86 ft above the transducer or 73 ft bgs. Recharge operations were curtailed later that day until the control system was fully operational.

Successful transmission of discharge and water level data to a computer at the FHWRP began on July 12, 2001. Hourly averaged flow rate and stage versus time are presented in Appendix F (see Figure F.2). Figure 2.14 presents the status of dry well operation throughout the data collection period. From July 12, 2001 through July 26, 2001, the dry well was operated continuously. Discharge and water level data were recorded every 10 seconds. The low and high settings were set at 69 ft and 109 ft above the transducer, respectively. This corresponds with a low setting of 90 ft bgs and a high setting of 50 ft bgs, respectively. When the control valve opened the flow rate typically spiked between 200 and 300 gpm while the recharge pipe filled. After 10 to 30 seconds the flow rate would stabilize to between 110 gpm and 140 gpm. The water level took approximately 2.0 to 2.5 minutes to reach the high setting. Upon reaching the high setting, the control valve would close until the water level receded to 90 ft bgs, which took between 10 and 15 minutes. During this period, the dry well was cycling on and off every 15 to 20 minutes.

The dry well was shut down on July 26, 2001 due to difficulties with the transducer cable and a lightening strike, which damaged both the instrumentation and control system and transducer. Turbulence in the dry well due to the flow surge, which occurred every time the flow control valve opened, caused the transducer cable to become severely entangled. To address this problem, EPWU replaced the 2-inch PVC transducer pipe with 2-inch PVC blank pipe and added vent holes to allow for air release. The transducer pipe was also strapped to the 4-inch recharge pipe to help maintain a more stable position. The transducer was raised 10 ft to 149 ft bgs and the low and high settings were changed to 39 ft and 139 ft above the transducer, respectively. This corresponds with a low setting of 110 ft bgs and a high setting of 10 ft bgs. The range between the high and low settings was expanded from 40 ft to 100 ft in an effort to reduce the frequency the well cycled on and off.

Recharge operations were resumed at the dry well on August 24, 2002 and continued until October 20, 2002. After modifying the low and high settings, the dry well cycled on and off every 25 to 55 minutes. In general, the water level in the well took approximately five to 10 minutes to reach the high setting. Upon reaching the high setting, the control valve would close until the water level receded to the low setting, which typically took between 20 and 45 minutes. The water level took progressively less time to reach the high setting and more time to recede as recharge operations continued.

The low setting for the dry well was reset on October 9, 2001. The low setting was raised from 39 ft above the transducer to 89 ft above the transducer in an effort to maintain water levels above the screened interval in the dry well. Previously, when the low setting was set at 110 ft bgs, the dry well screen was continually exposed to air every time the flow control valve closed and the water level receded to the low setting. By raising the low setting, the water level could not drop below the top of screen before the flow control valve opened. After raising the low setting, the well cycled on and off every 15 to 25 minutes. The water level took much less time to reach the high setting and then recede because the range between the low and high setting was reduced from 100 ft to 50 ft.

The flow to the dry well was subsequently reduced to an average of 70 to 75 gpm in an effort to extend the injection period (on/off cycle) or eliminate injection cycling altogether. A recharge rate that would sustain a near constant water level in the well was desirable because the volume of air being forced into the well and formation would be reduced significantly. After adjusting the flow rate to the well the cycling time increased marginally, although the increase in cycling time was much less than predicted. Attempts to reduce the flow rate further were unsuccessful due to the size and type of PRV and valve installed at the reclaimed water pipeline.

The well was shut down on October 20, 2001 due to battery failure. Recharge operations were restarted at the well on October 30, 2001 and continued through November 8, 2001. During that time the dry well cycled on and off every 20 to 30 minutes. In general, the water level in the well took approximately three to four minutes to reach the high setting and approximately 15 to 20 minutes to recede to the low setting.

On November 8, 2001 recharge operations at the dry well were suspended until March 3, 2002 due to a variety of equipment problems, maintenance, and lack of supply from the FHWRP. Problems encountered during this period are described in more detail under the section titled "Operational Difficulties and Potential Improvements."

Recharge operations were resumed on March 3, 2002 and continued until April 25, 2002. Upon restarting recharge operations, the performance of the well appeared to have deteriorated based on the water level and flow data recorded. The average flow rate into the well still ranged from 70 to 75 gpm, however, the water level took two to three minutes to reach the high setting and about two to three hours to recede to the low setting.

The poor performance of the well from March 2002 through April 2002 may be due to a number of factors. The water level and flow data collected may be erroneous. The meter used to record flows may have malfunctioned as a result of rust, corrosion, or chemical build-up due to a long period of non-operation. The poor performance may also be a function of the well itself. For example, chemical scale or bio-fouling may be present through the screened interval, impeding movement of water from the well into the surrounding formation. The accumulation and/or trapping of air in the injection interval may also be a cause of poor performance, however, it does not explain the marked drop in performance from November to March. Due to possible inaccuracies associated with the data collected from March through April, this data was not used in developing any conclusions as to the overall performance of the well.

Recharge operations at the dry well were essentially suspended from May 2002 through July 2002 due to a lack of available water from the FHWRP.

MONITORING METHODS AND DATA COLLECTION (TASK 6)

Water infiltration through the vadose zone to the water table was monitored using the following methods:

- Automated monitoring of thermocouple temperature profiles beneath the basin floor
- Periodic water level measurements in monitor wells and vadose zone piezometers
- Collection and lab analysis of water quality samples from monitor wells and vadose zone piezometers
- Neutron moisture logging of monitor wells

Thermocouple Installation and Monitoring

On July 9 and 10, 2001 an array of thermocouples was installed beneath the infiltration basin floor. The thermocouples were installed to monitor diurnal temperature fluctuations at multiple shallow depths beneath the basin floor. The El Paso, Texas area experiences large changes in air temperature from day to night. Coupled with the fact that the infiltration basin was designed for shallow water depths, the pool of water in the basin warms up and cools down significantly throughout the course of a day. During periods of infiltration, alternating pulses of warmer and cooler water move downward beneath the basin. By monitoring the daily (diurnal) temperature fluctuations at multiple depths beneath the basin, the time lag, and rate of attenuation in amplitude with depth of the temperature waves can be used to estimate rates of water infiltration beneath different portions of the basin. Subsurface temperature profiling to estimate water recharge rates beneath streams and ponds has been performed previously by other researchers (e.g. Constantz et al. 1999, Taniguchi 1993). Stallman (1965) has shown that relatively high infiltration rates (>2 cm/day) can be monitored by analyzing diurnal subsurface temperature fluctuations, whereas estimation of slower infiltration rates (<2 cm/day) require analysis of longer period seasonal or annual subsurface temperature data.

The depths of thermocouple placement were chosen to optimize observation of diurnal temperature waves, which generally die out at depths of a few meters below surface (e.g. Thomas et al. 2000, Lapham 1989). The depths of thermocouple placement were considered optimal for this study. While seasonal and yearly temperature fluctuations may be observed to much greater depths, these longer period temperature fluctuations are of less interest to this study, which was aimed primarily at quantifying changes in seepage rates through the floor of the basin as a result of a surficial clogging layer. In any case, additional equipment would have been necessary to advance borings or drivepoints to deeper depths (beyond one meter) because hand auguring is often limited in such materials to approximately a meter in depth. The associated costs of installing deeper thermocouples was beyond the scope of this project.

Thermocouples were installed at five locations (Stations A, B, C, D, and E) along a north-south transect across the basin floor. At each station, thermocouples were installed at three depths (30 cm, 60 cm, and 90 cm below surface). Each thermocouple consisted of a bimetal junction (Omega Engineering ANSI Type T) capable of measuring subsurface temperatures to ±0.1°C. A hand auger was used to drill 1-meter deep boreholes into the loose sand and gravel that underlies the basin floor. In addition to the 15 subsurface thermocouples, two additional thermocouples were installed near the basin floor at Stations A and D to monitor pool water temperatures within the infiltration basin.

Thermocouple wires were routed through a 1.5-inch diameter polyethylene tube extending to the north edge of the basin, where the wires were connected to a Campbell 21X Micrologger. The data logger was programmed to collect temperature measurements every 10 minutes. This frequency of temperature measurements is the same as that employed by Thomas et al. (2000). An additional thermocouple was placed above the water level to monitor air temperature at the edge of the infiltration basin.

The system was tested to ensure that all of the thermocouples were functioning properly, and began collecting temperature data on the morning of July 10, 2001. Data was collected from 7/10/01 to 5/15/02 and is presented in Appendix G for each thermocouple. Data is missing from 7/21/01 to 8/7/01 and from 8/22/01 through 10/13/02. For the data set from 3/22/02 to 5/15/02, several outlying series of data points were eliminated: (1) A90: Erroneously high values (above 40°C) after 5/15/02 3:04 pm were removed; (2) B30: Both high and low values (above 40°C or below 20°C) after 5/15/02 3:32 pm were removed; (3) B60: Erroneously low values (below 20°C) after 5/15/02 6:04 pm were removed; (4) C30: Erroneously high values (above 40°C) after 5/15/02 1:34 pm were removed; and (5) C60: Erroneously low values (below 20°C) after 5/15/02

2:48 pm were removed. It was assumed that these anomalous readings were due to disturbance of the thermocouples, perhaps during basin cleaning.

Water Level Measurements

Water level measurements in monitor wells M1 and M4 were made manually using an electric sounder. Depths to water are provided in Table 2.3. In addition, water level dataloggers were installed temporarily in piezometers M4/P2 (10/30/01 to 11/15/01) and M4/P3 (7/30/01 to 10/11/01) to record the buildup and dissipation of perched water in the vadose zone. Water level measurements in piezometers were also made manually. Hydrographs for M1 and M4 and piezometer M4/P3 are included in Figures 2.15 and 2.16, respectively.

Water Quality Sampling

Following the measurement of static groundwater levels, groundwater samples were collected using a ½-gallon PVC bailer and motorized winch powered by a 12-volt car battery.

Table 2.3

Summary of water level measurements

Date	Water level (ft below top of casing)	
	M1	M4
07/16/01	393.30	386.63
07/18/01	393.27	386.55
07/25/01	393.29	386.68
08/02/01	393.27	386.72
08/07/01	393.37	386.81
08/16/01	392.98	386.10
08/23/01	393.21	385.40
09/04/01	393.33	384.43
09/10/01	392.85	384.21
09/20/01	392.19	384.34
09/25/01	392.12	384.19
10/02/01	391.93	383.97
10/16/01	391.90	384.21
10/25/01	393.10	383.93

The bailer was run up and down the 400-foot deep monitor wells using a portable tripod and pulley. Prior to sample collection, approximately three casing volumes of groundwater were purged from each well, or until the field parameters, which included temperature, pH, and EC, became stable. Water quality samples were collected by EPWU staff in appropriate containers, and were hand delivered to the EPWU laboratory for analysis. Water quality laboratory results and field parameter measurements are summarized in Tables 2.4 and 2.5, respectively.

Neutron Logging

Neutron logging was performed using a Campbell Pacific Nuclear Hydroprobe (Serial No. 38018058). A standard count was performed prior to logging each monitor well, and the Hydroprobe automatically calculates chi-squared statistics to ensure proper functioning of the electronics. Logging of monitor wells was performed at 2-foot depth intervals using a 32-second count interval down to a maximum depth of 300 ft, which was the length of the longest probe cable available. Raw count values were recorded in the field logbook, and subsequently entered into an Excel spreadsheet to correct depths to ground level and for plotting of neutron logs. Neutron logging was completed on July 10, July 12, July 30-31, and October 8-10, 2001. Neutron logs for monitor wells M1, M3, and M4 are shown in Figure 2.17.

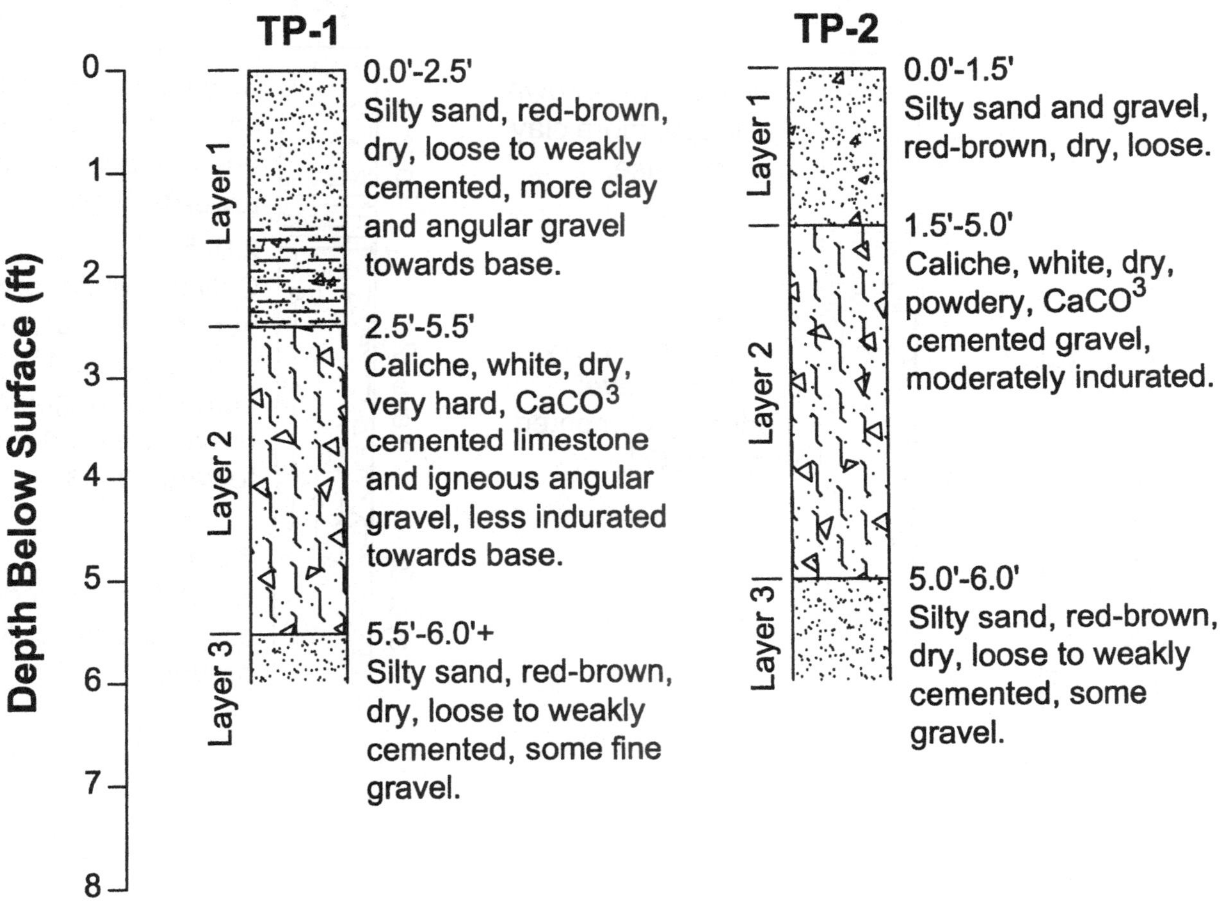

Note: Test pits near City of El Paso Well 41 excavated to six feet deep using trackhoe excavator.

Figure 2.3 Test pit stratigraphic sections

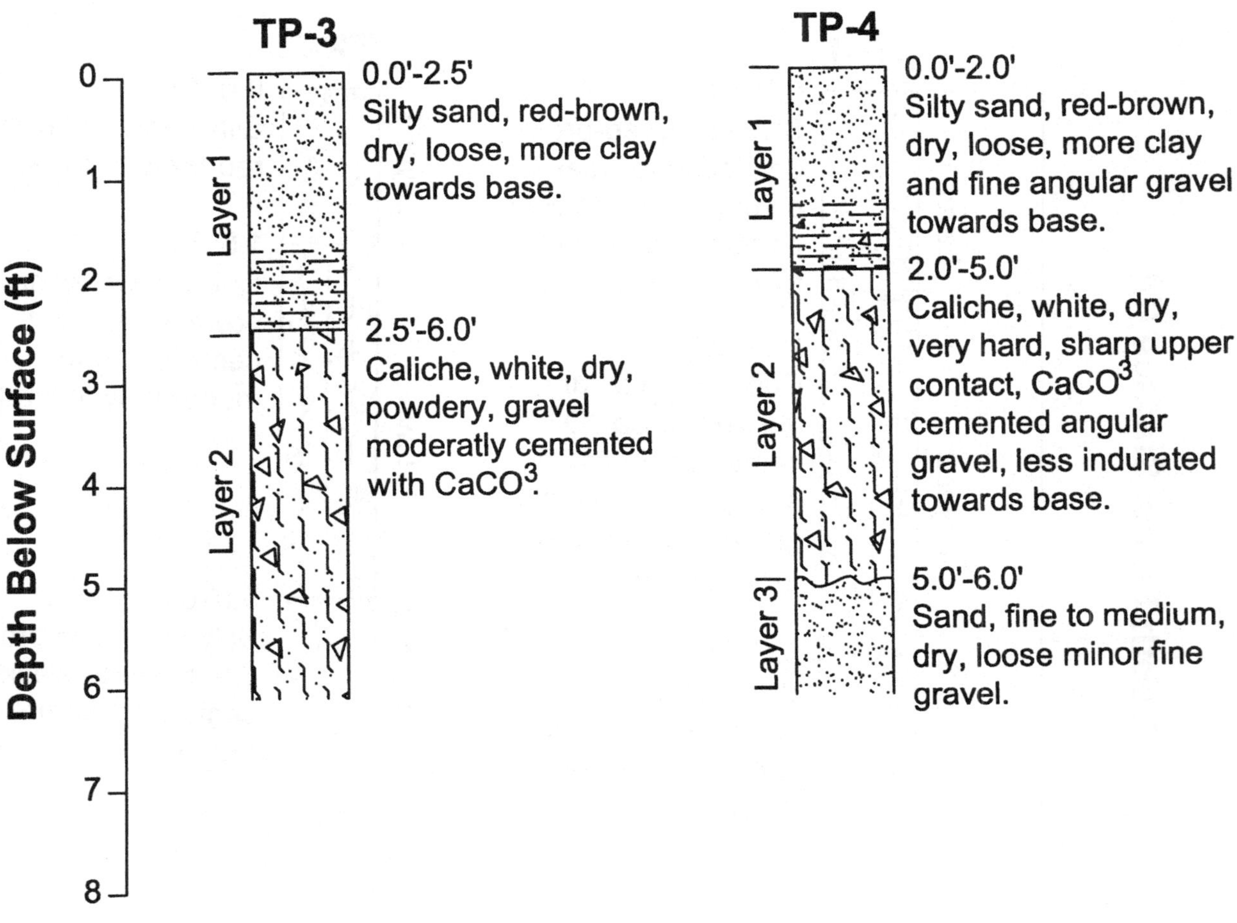

Note: Test pits near City of El Paso Well 41 excavated to six feet deep using trackhoe excavator.

Figure 2.3 (continued) Test pit stratigraphic sections

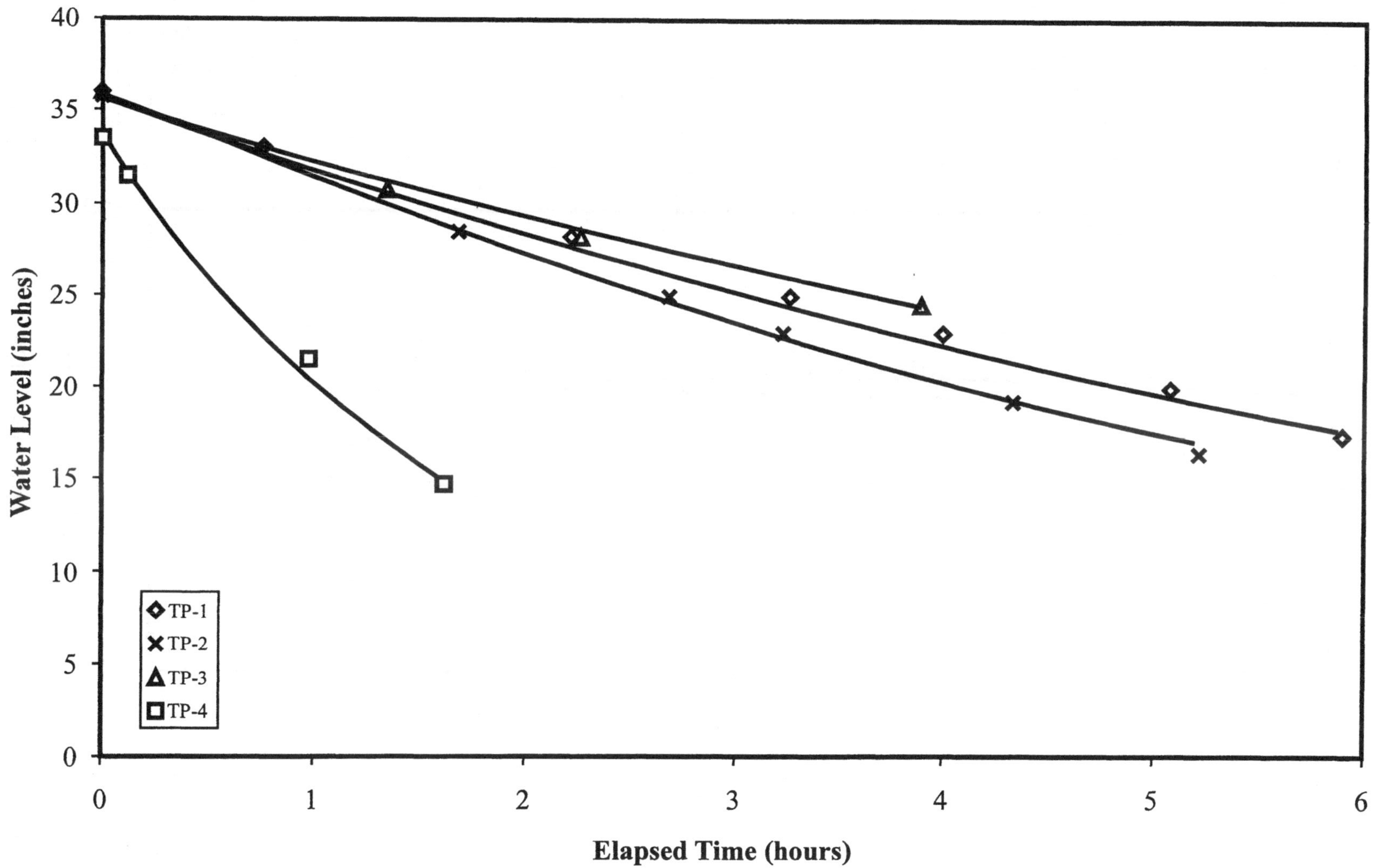

Figure 2.4 Test pit percolation rates

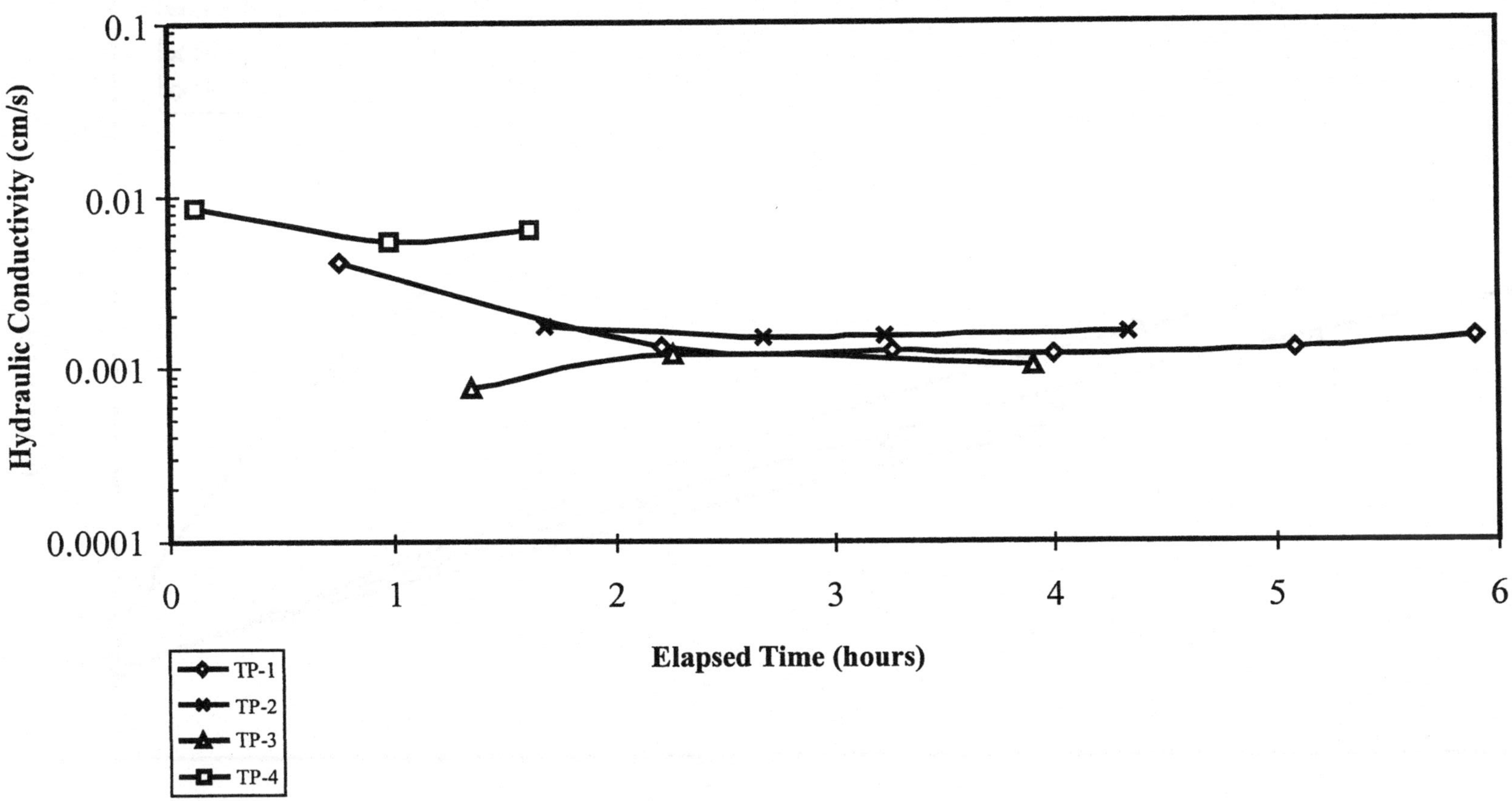

Figure 2.5 Test pit calculated hydraulic conductivity

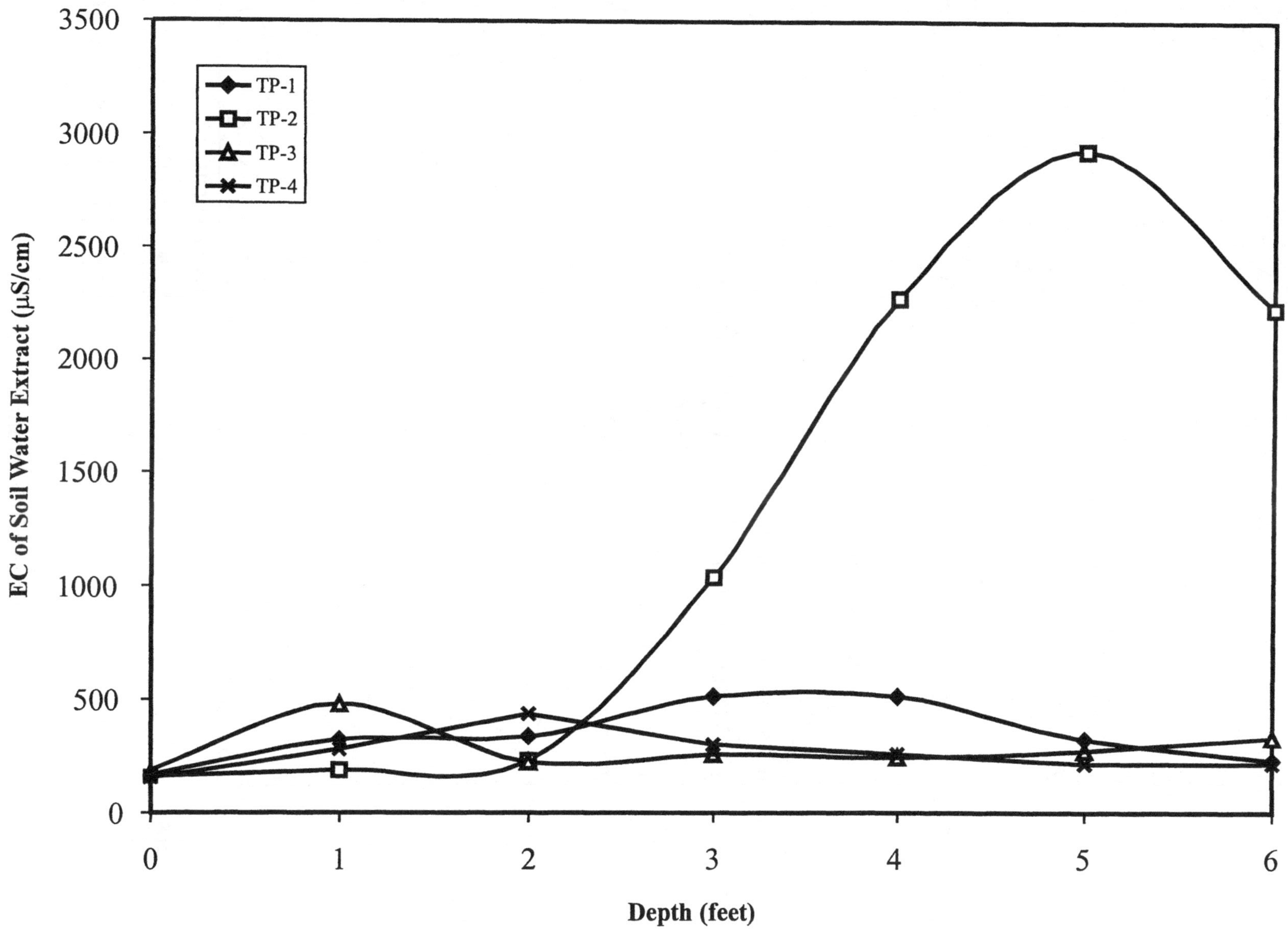

Figure 2.6 Test pit soil electrical conductivity results

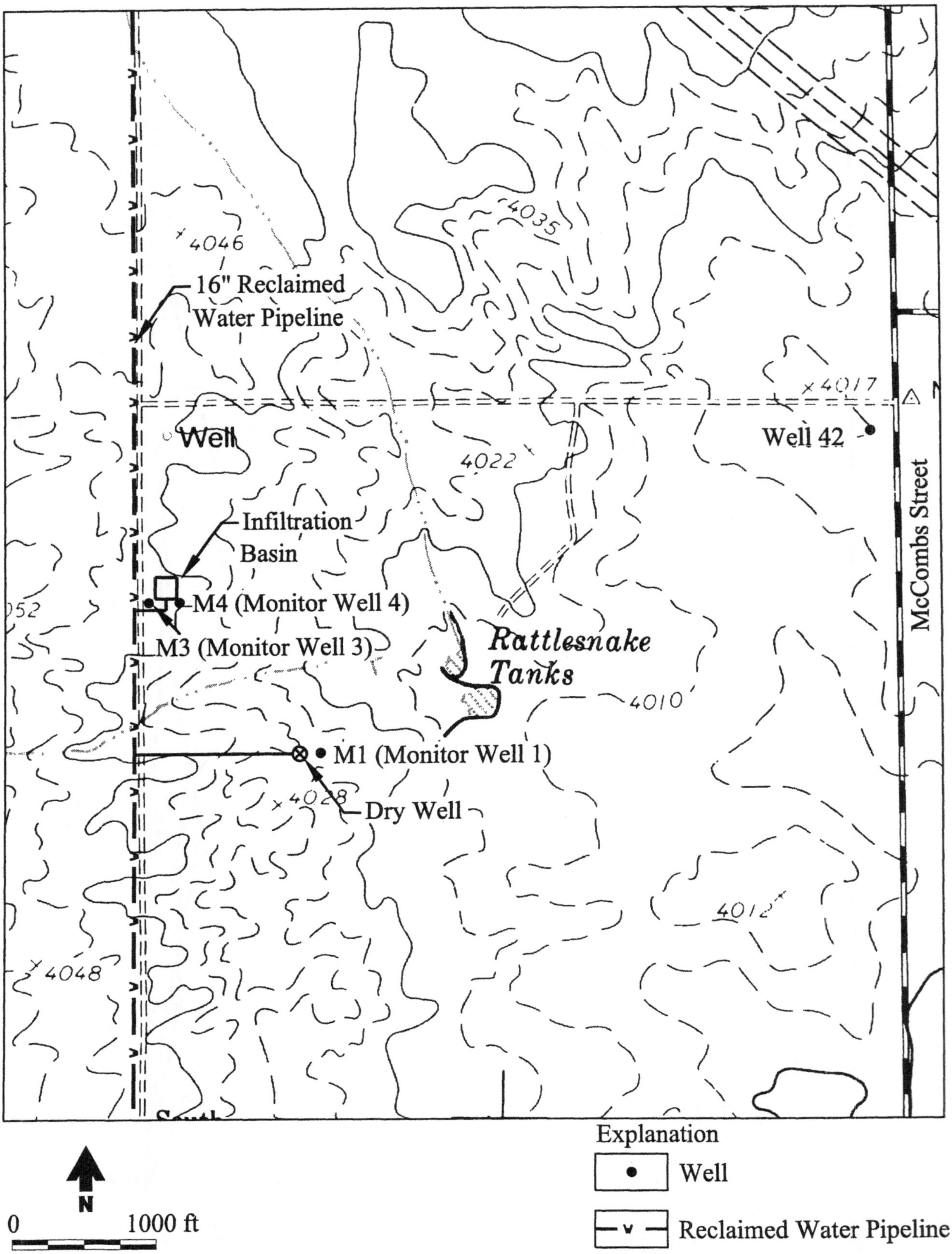

Figure 2.7 Approximate recharge facility locations

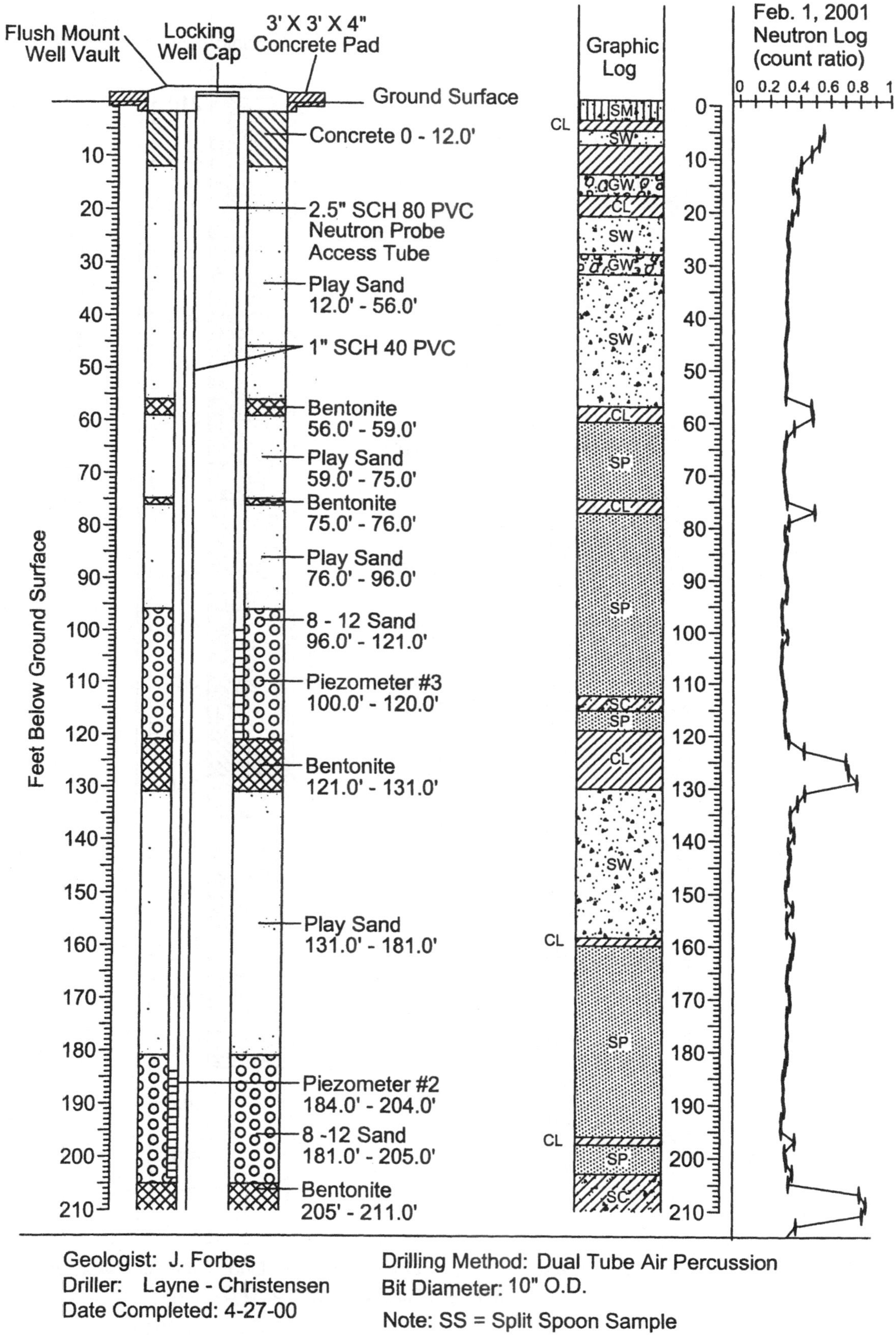

Figure 2.8 Monitor well M1 completion and lithologic log

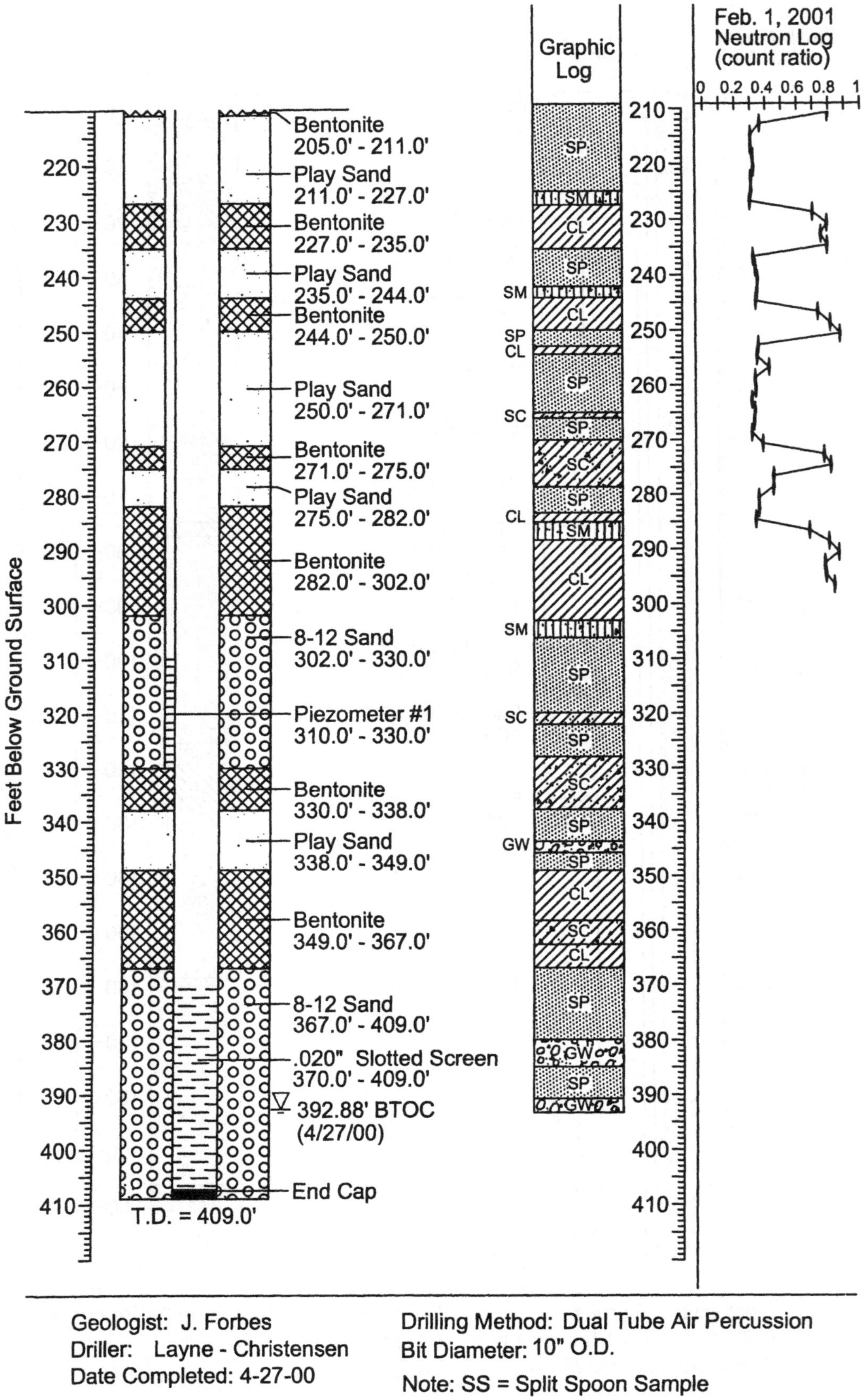

Figure 2.8 (continued) Monitor well M1 completion and lithologic log

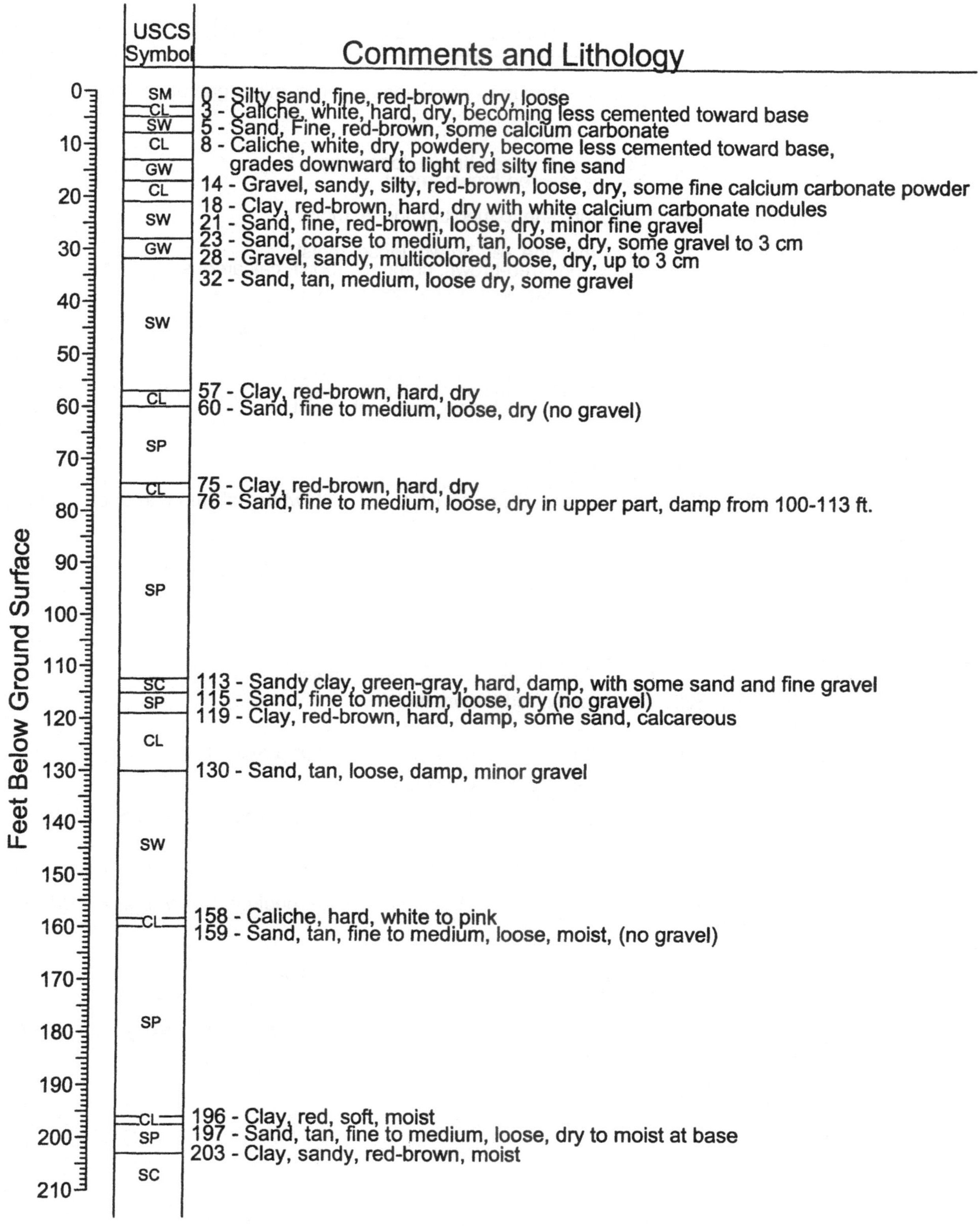

Geologist: J. Forbes
Driller: Layne - Christensen
Date Completed: 4-27-00

Drilling Method: Dual Tube Air Percussion
Bit Diameter: 10" O.D.
Note: SS = Split Spoon Sample

Figure 2.8 (continued) Monitor well M1 completion and lithologic log

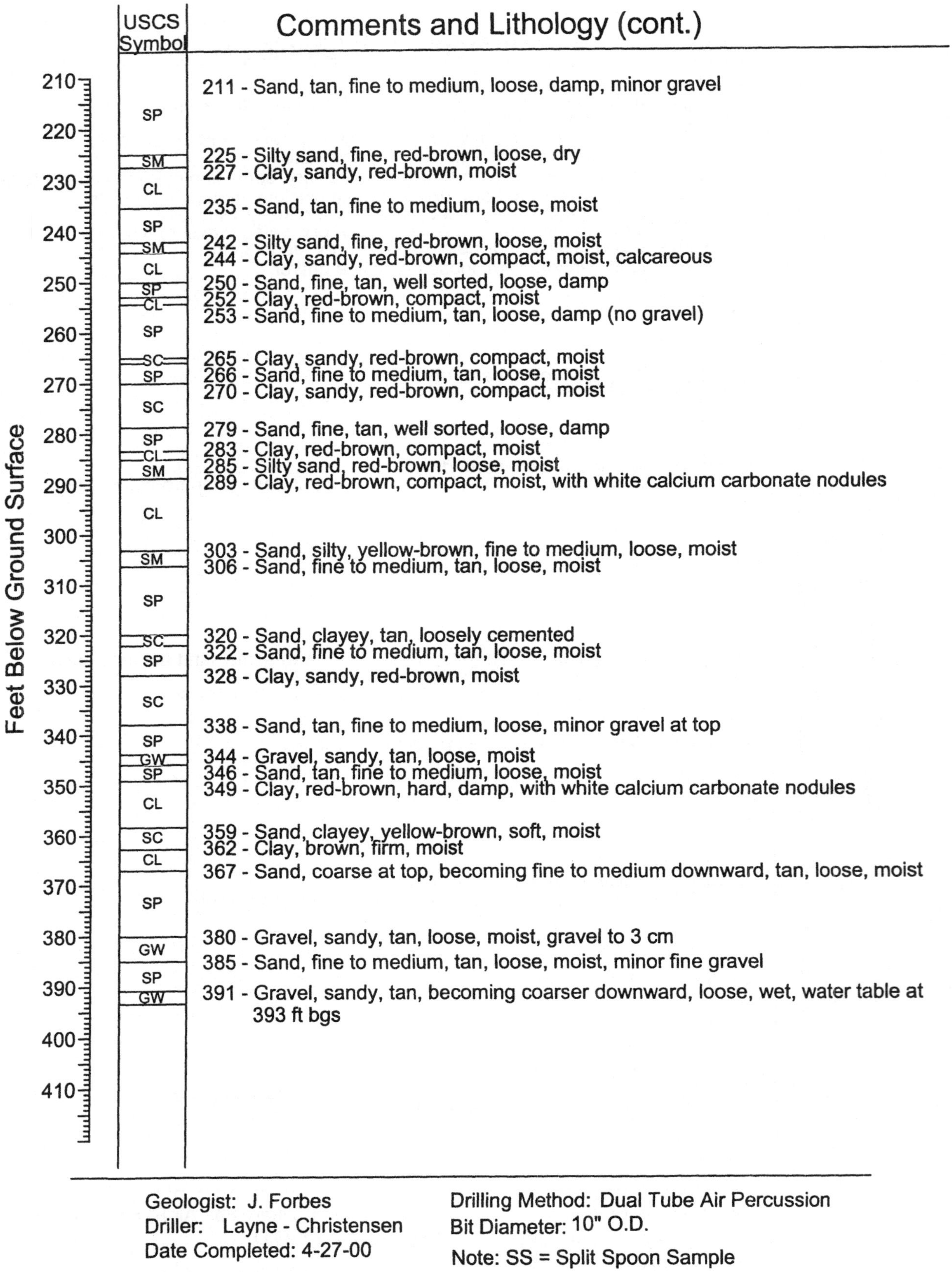

Figure 2.8 (continued) Monitor well M1 completion and lithologic log

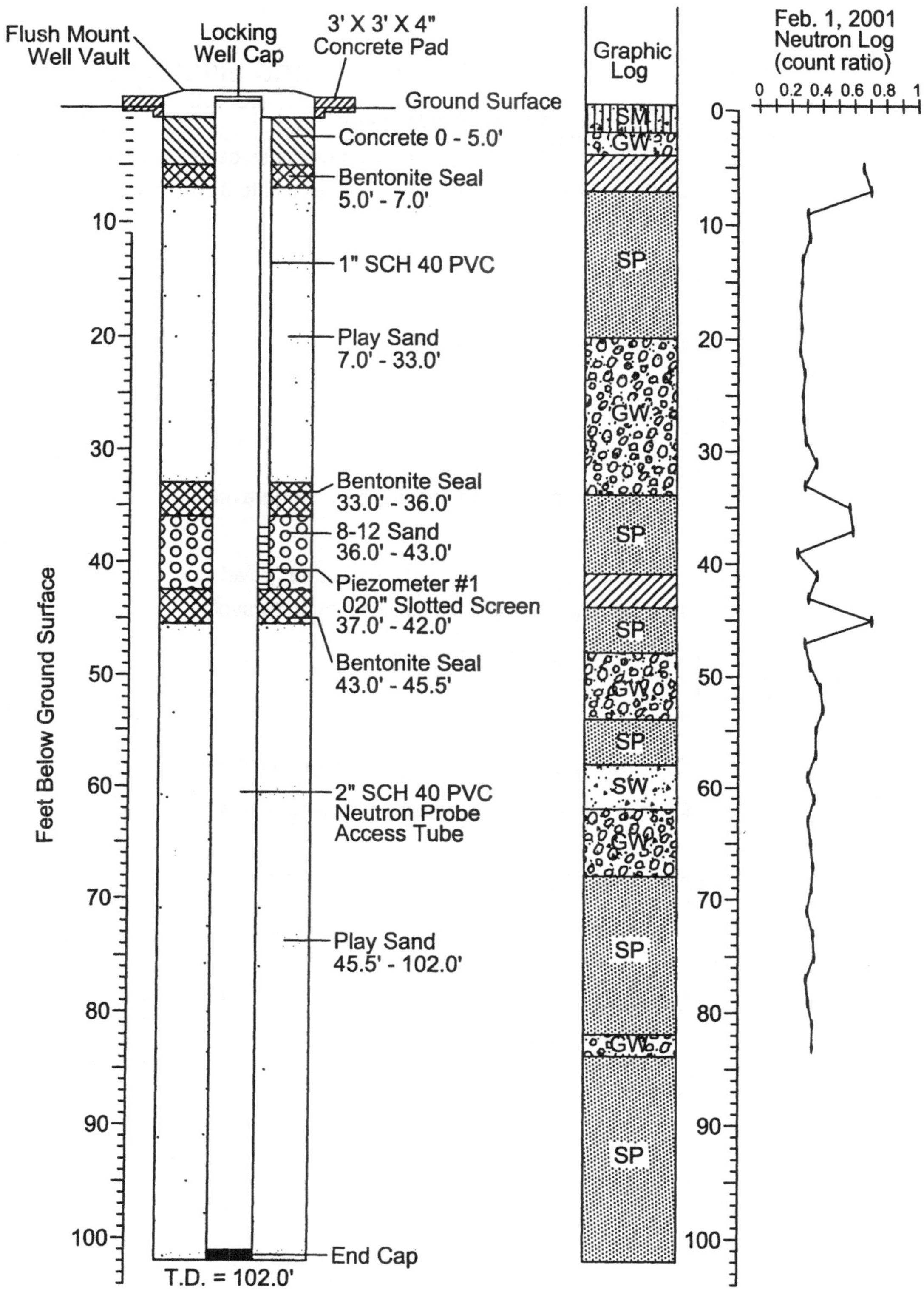

Geologist: J. Forbes
Driller: Layne - Christensen
Date Completed: 4-19-00

Drilling Method: Dual Tube Air Percussion
Bit Diameter: 10" O.D.
Note: SS = Split Spoon Sample

Figure 2.9 Monitor well M3 completion and lithologic log

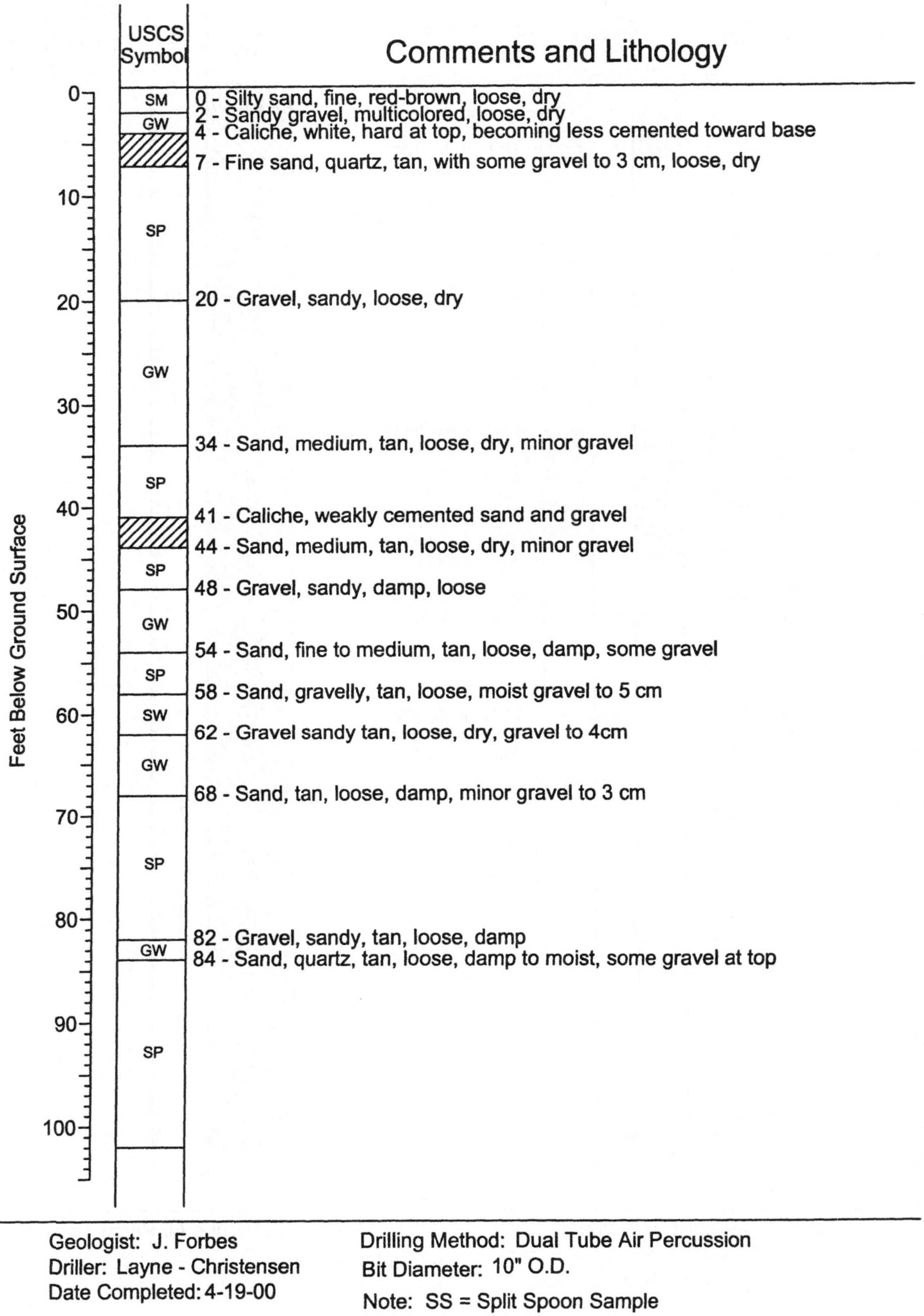

Figure 2.9 (continued) Monitor well M3 completion and lithologic log

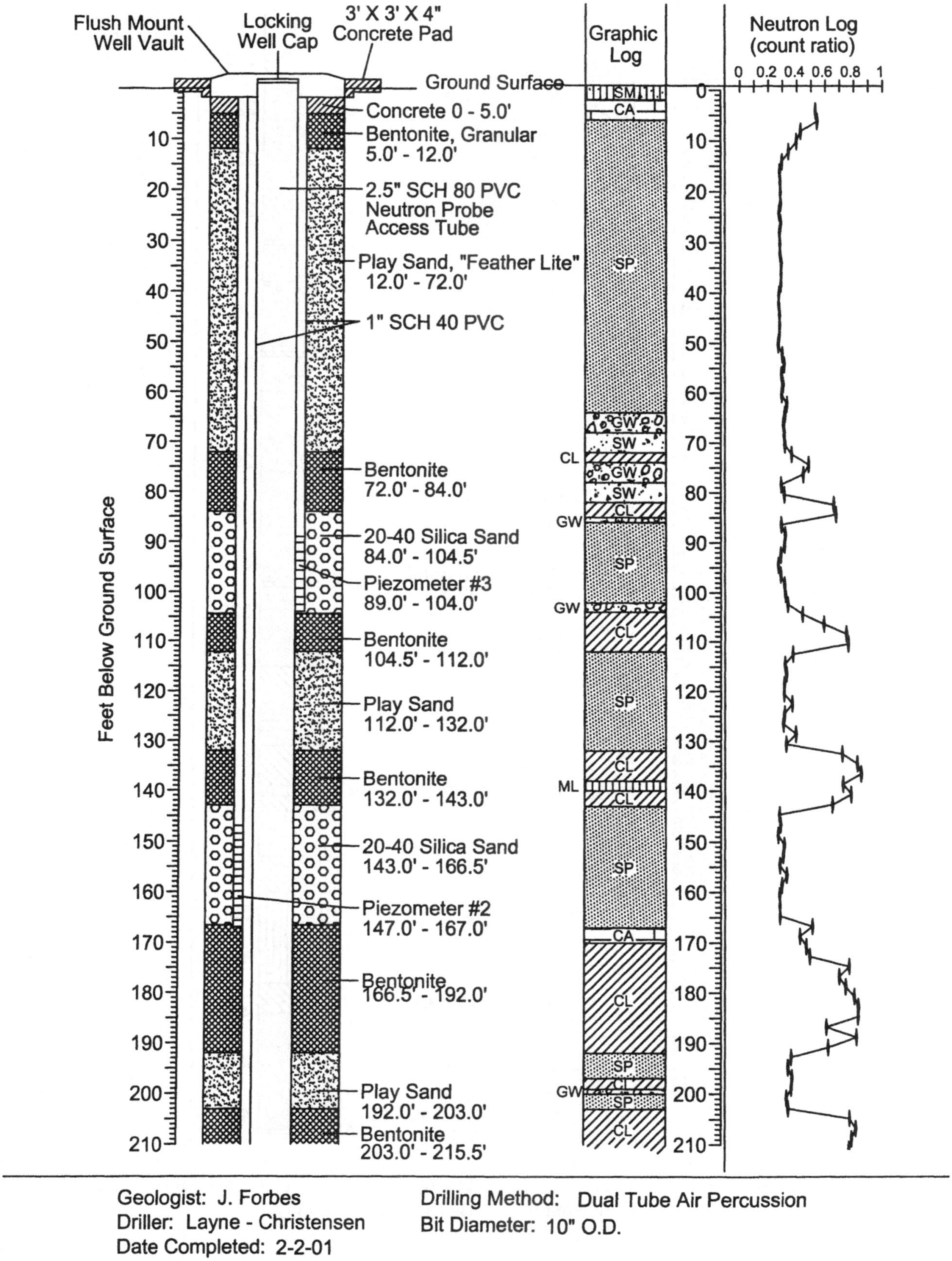

Figure 2.10 Monitor well M4 completion and lithologic log

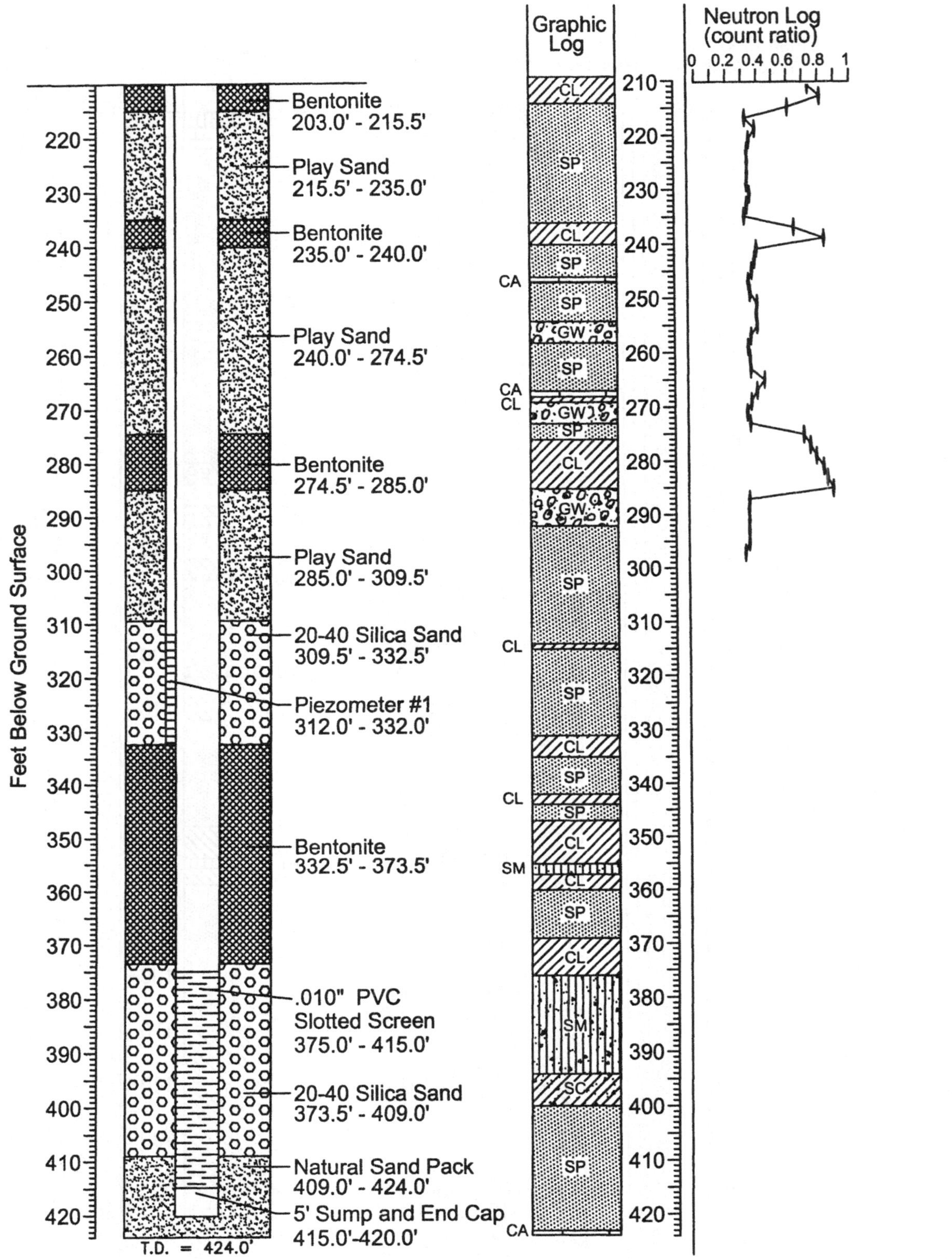

Geologist: J. Forbes
Driller: Layne - Christensen
Date Completed: 2-2-01

Drilling Method: Dual Tube Air Percussion
Bit Diameter: 10" O.D.

Figure 2.10 (continued) Monitor well M4 completion and lithologic log

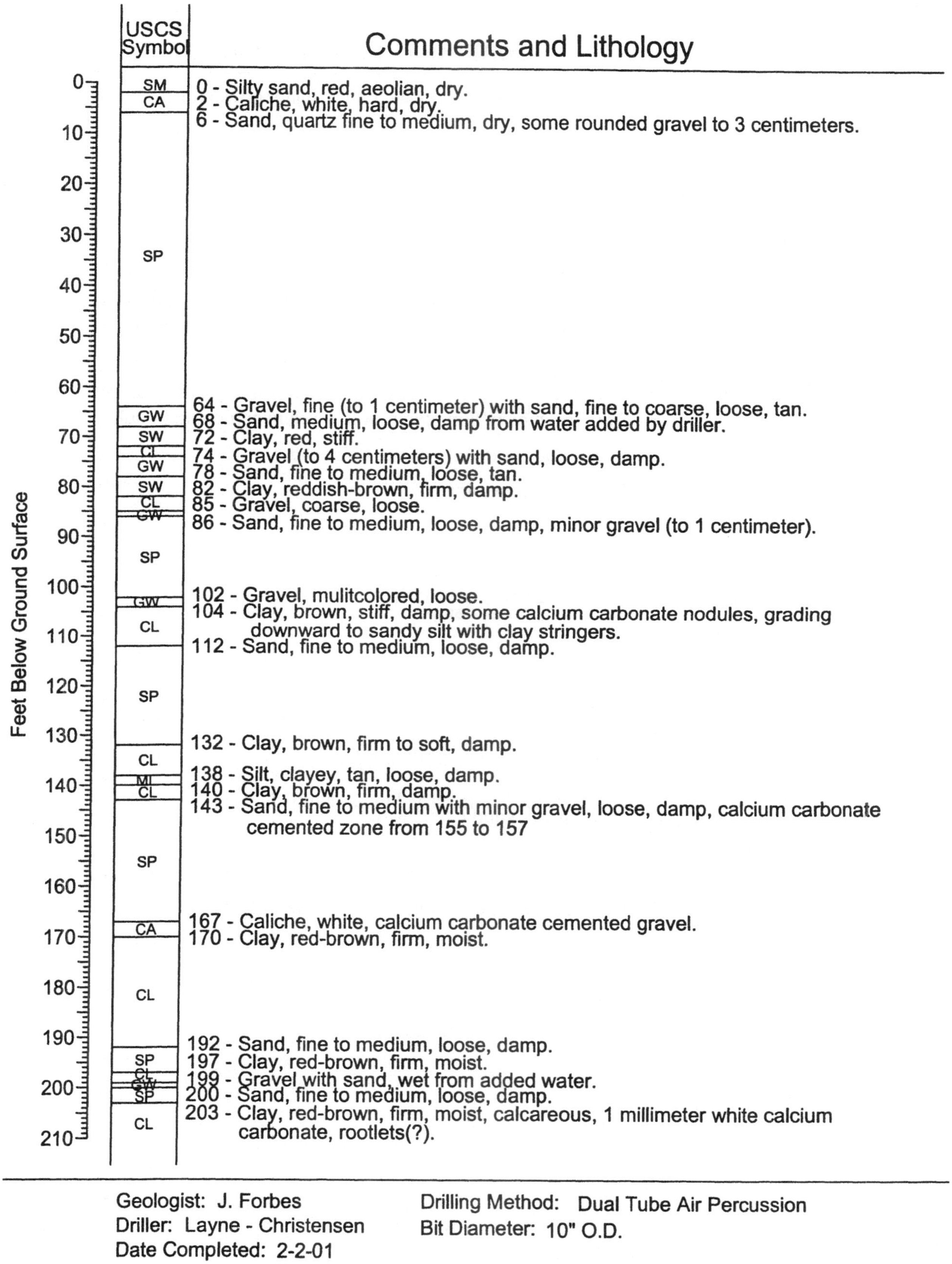

Figure 2.10 (continued) Monitor well M4 completion and lithologic log

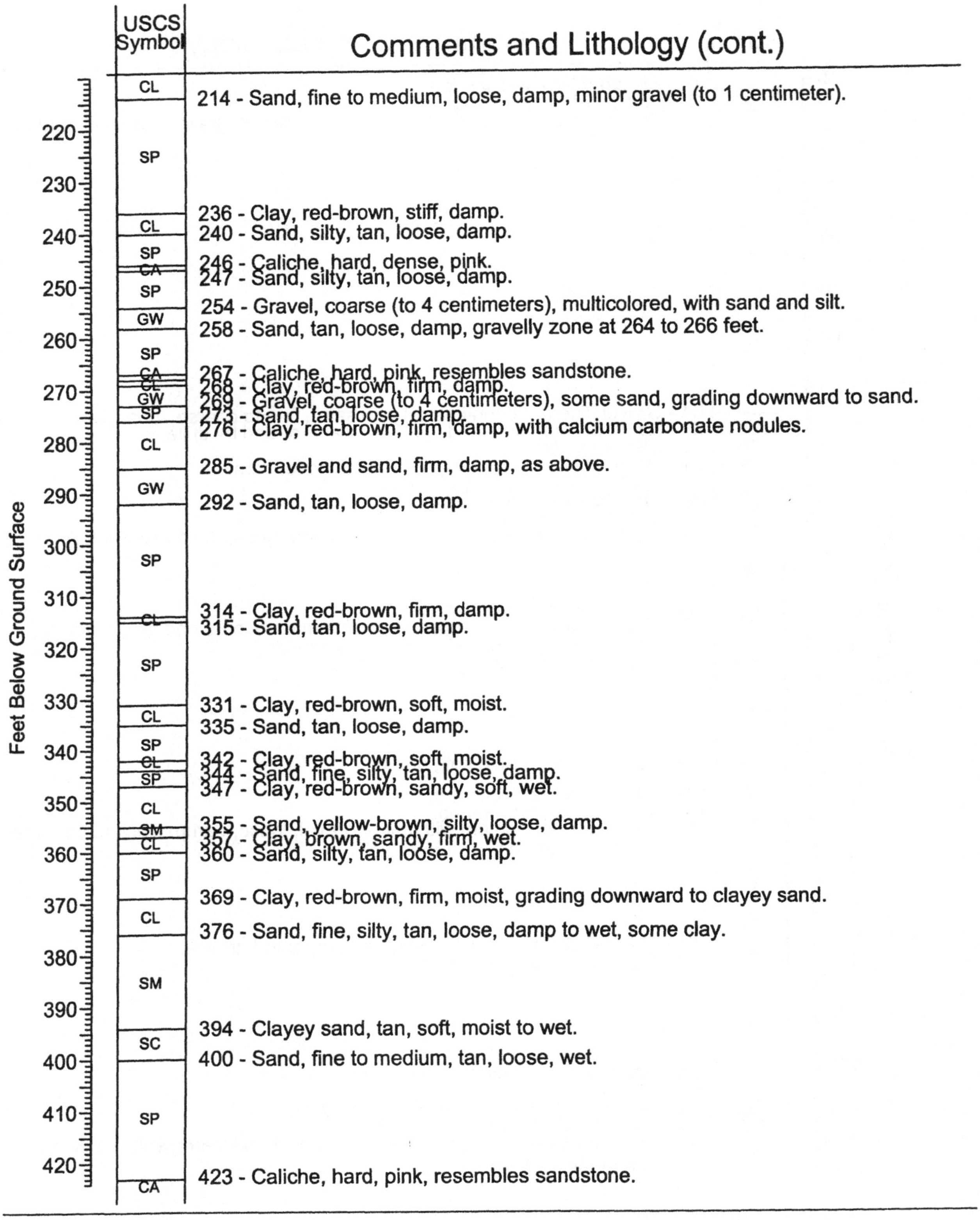

Geologist: J. Forbes
Driller: Layne - Christensen
Date Completed: 2-2-01

Drilling Method: Dual Tube Air Percussion
Bit Diameter: 10" O.D.

Figure 2.10 (continued) Monitor well M4 completion and lithologic log

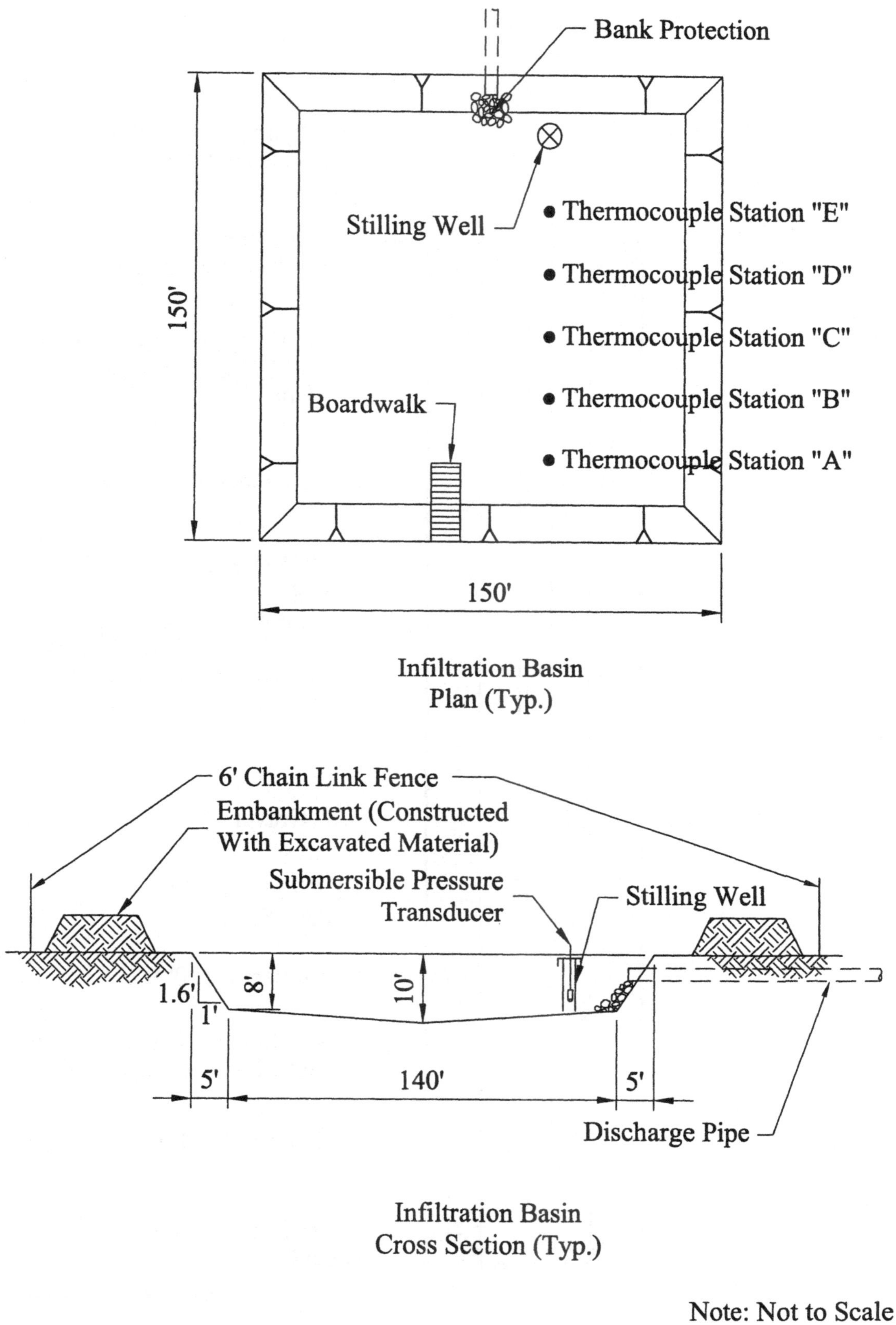

Figure 2.11 Infiltration basin as-built drawing

Engineer/Geologist: H.Thompson, J.Forbes
Driller: Layne-Christensen
Date Completed: 2-5-01
Drilling Method: Cable Tool Rig
Drive Casing: 20"

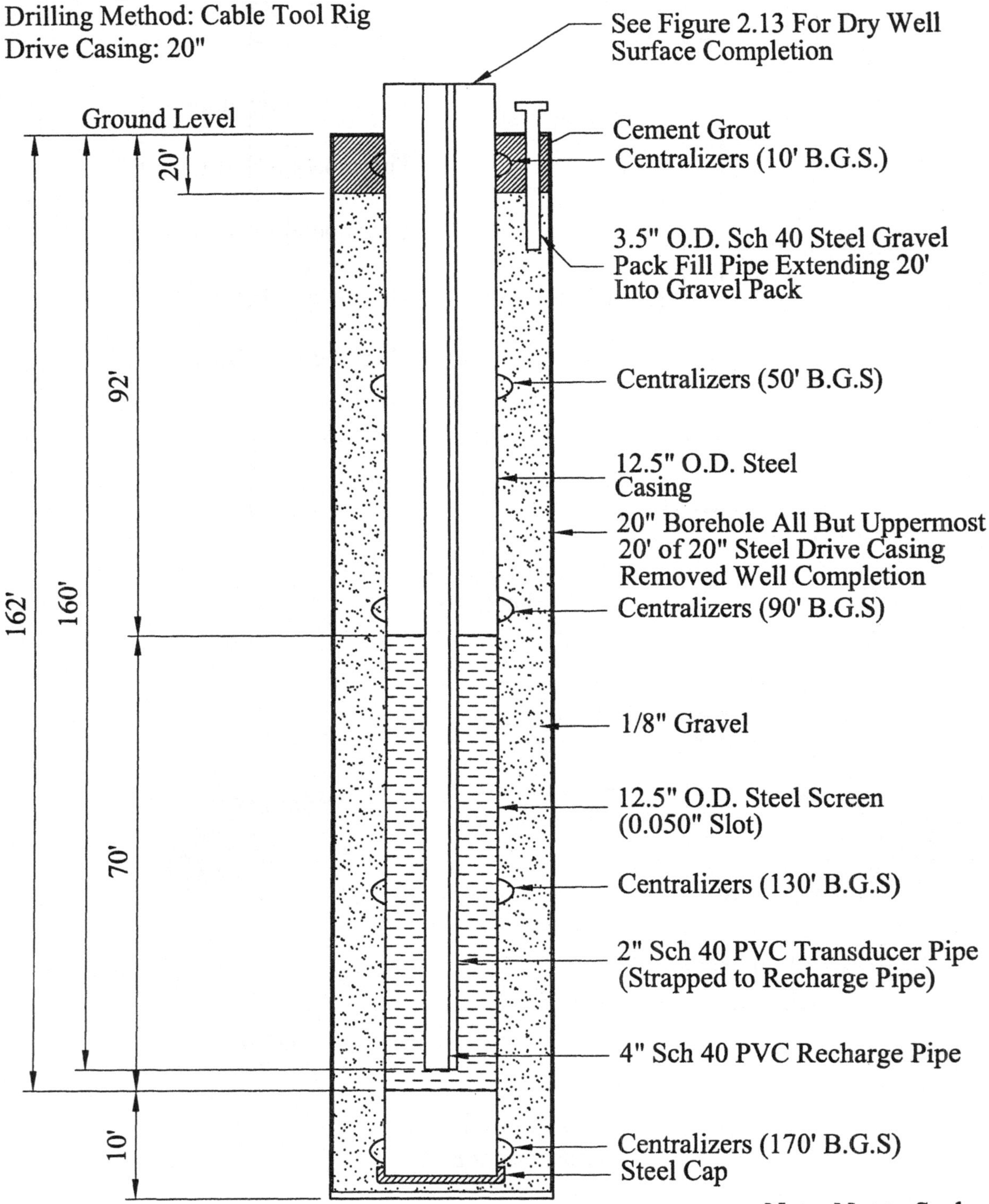

Note: Not to Scale

Figure 2.12 Dry well as-built drawing

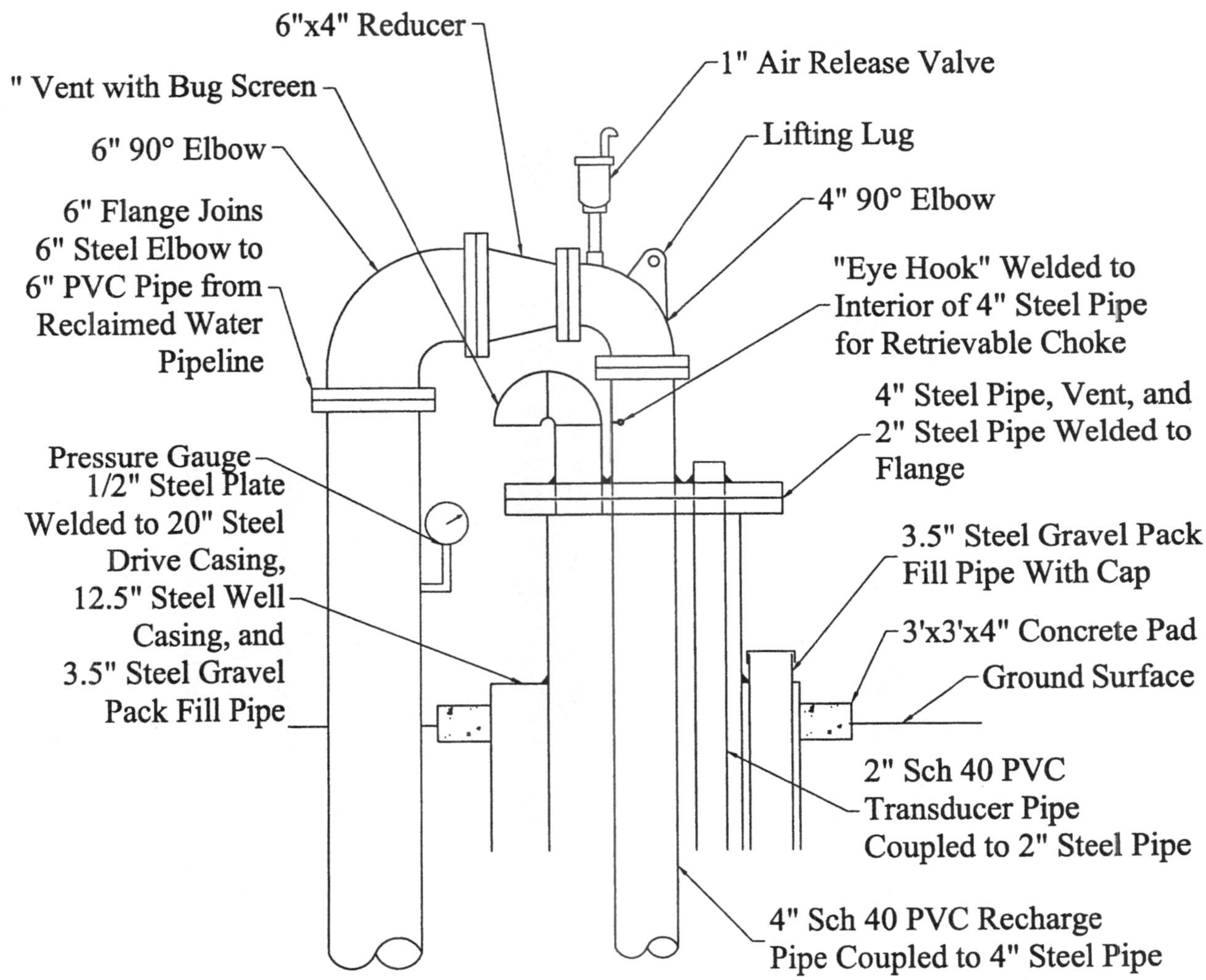

Figure 2.13 Dry well surface completion

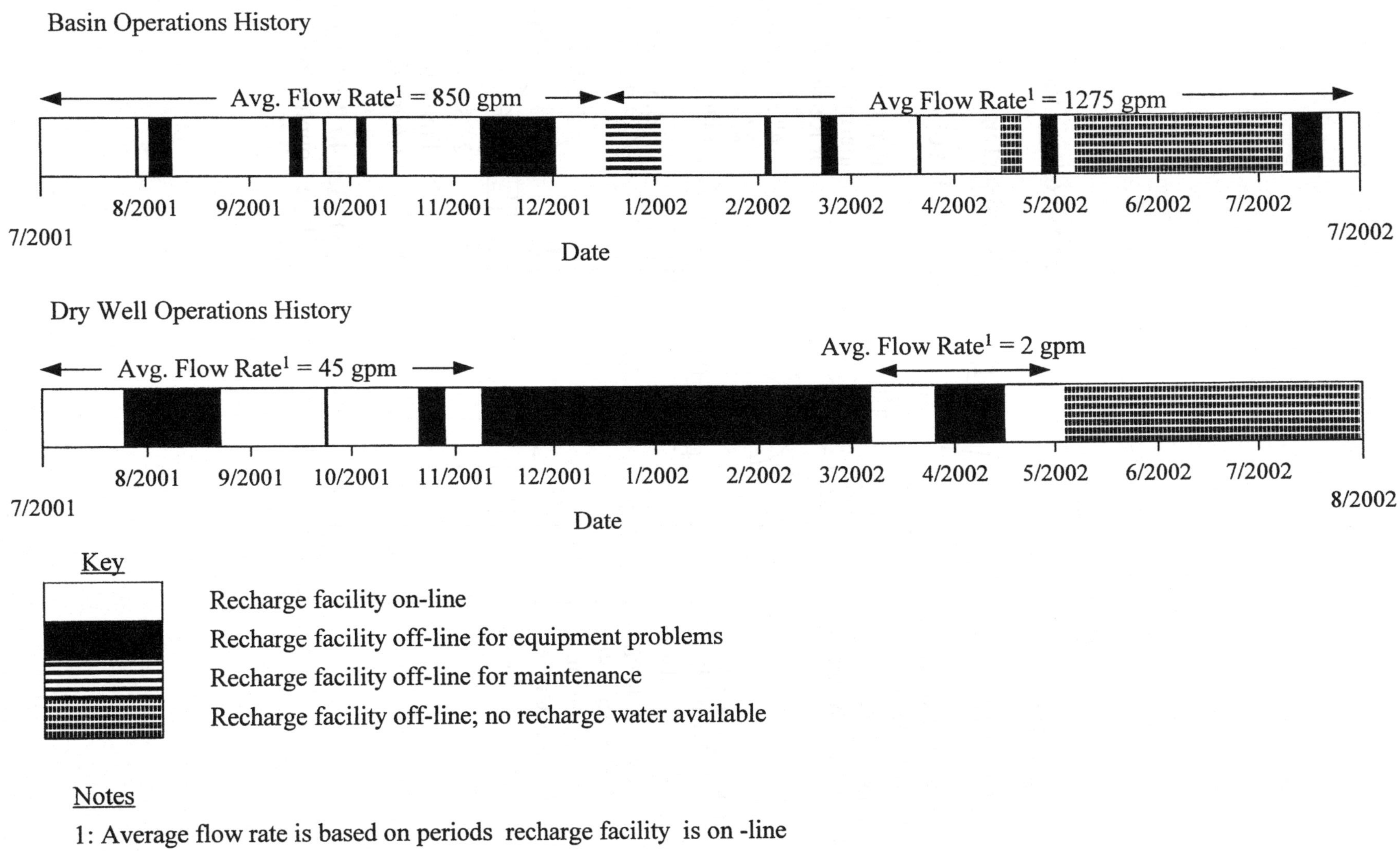

Figure 2.14 Basin and dry well operations history

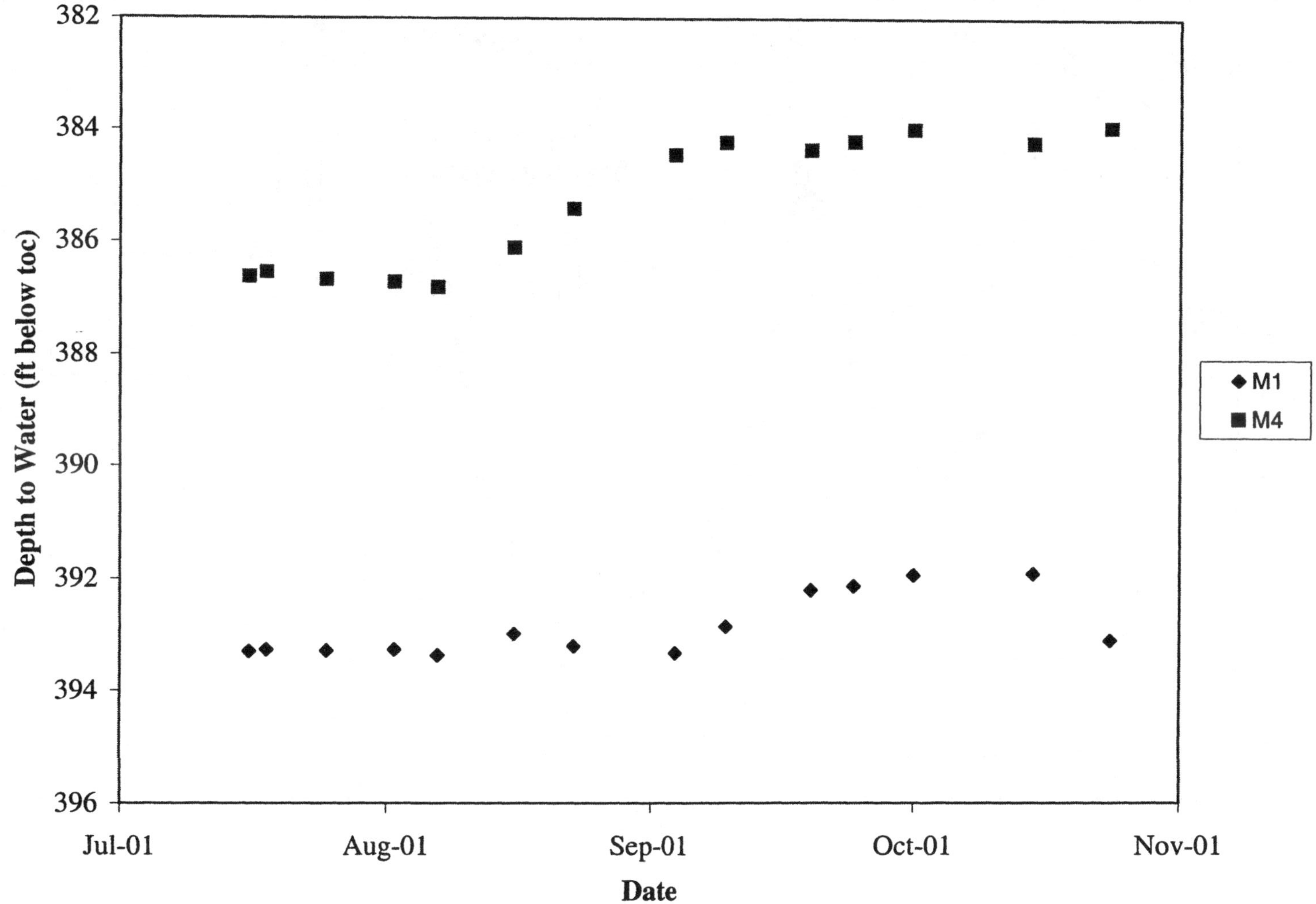

Figure 2.15 Water levels in M1 and M4

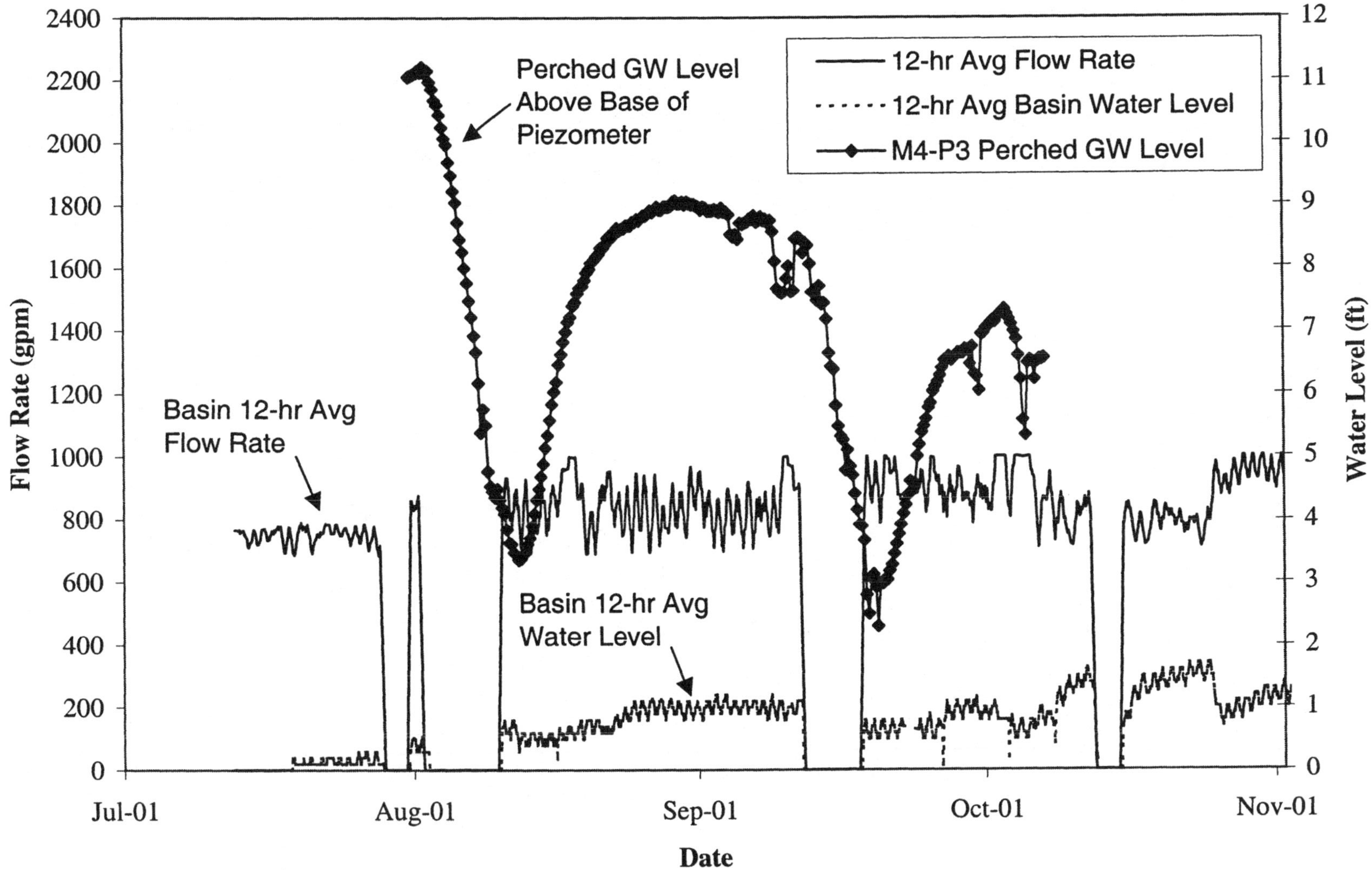

Figure 2.16 Perched groundwater buildup at M4/P3

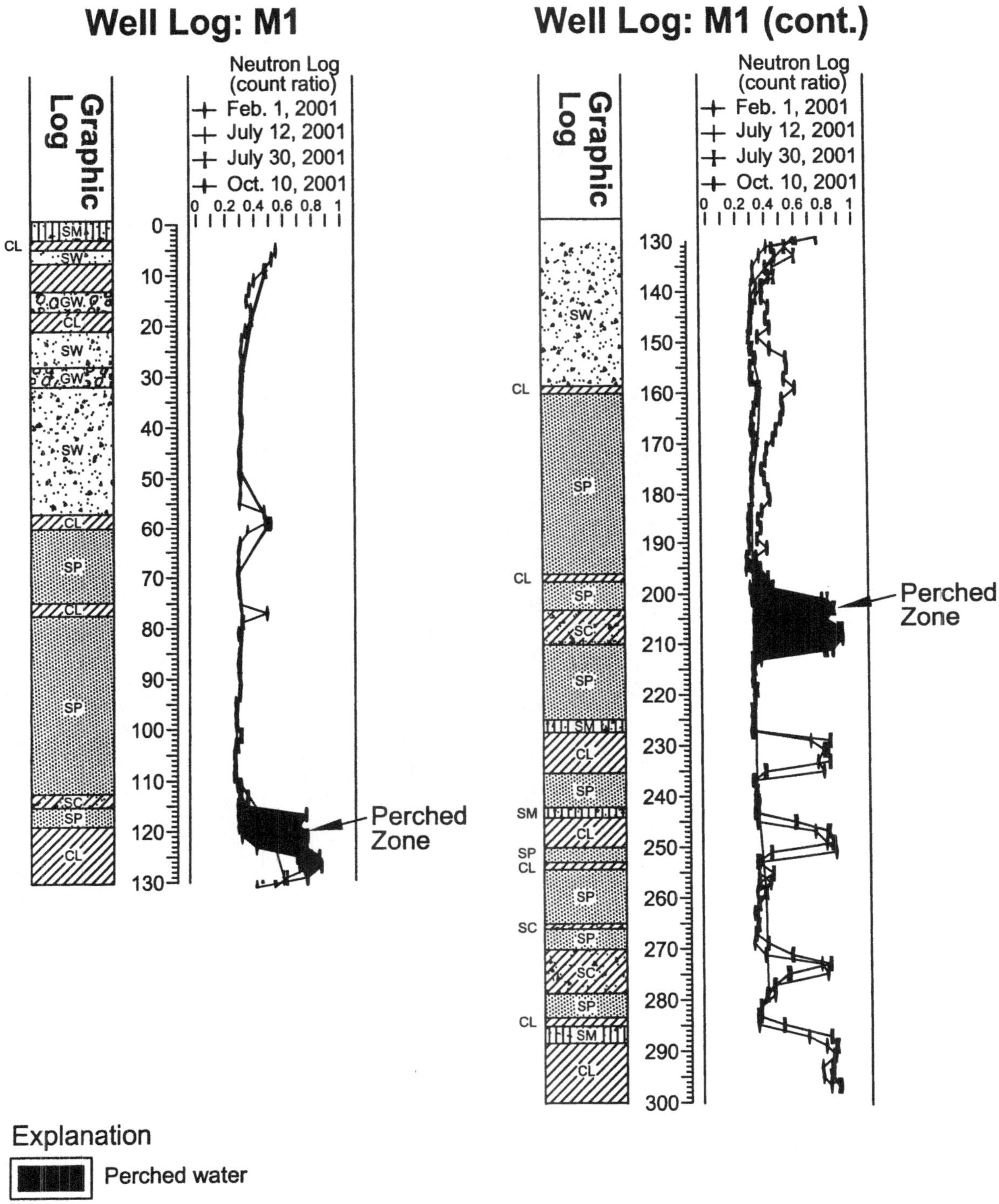

Figure 2.17 Neutron moisture logs showing perched water zones

Well Log: M3

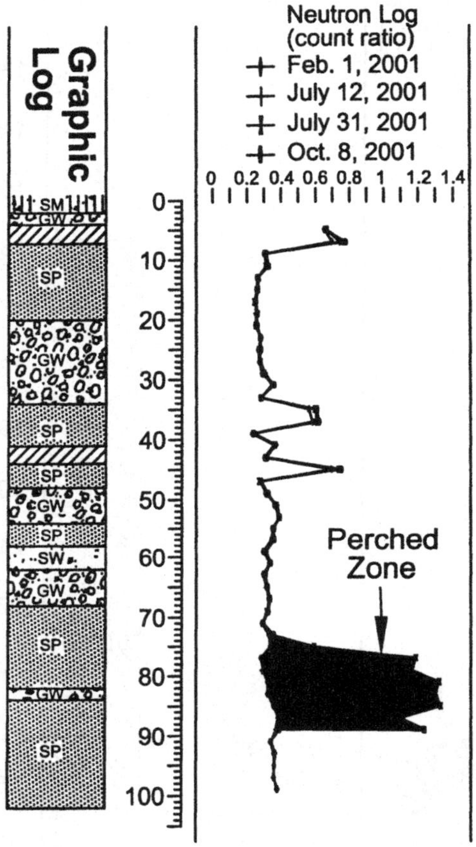

Explanation

Perched water

Figure 2.17 (continued) Neutron moisture logs showing perched water zones

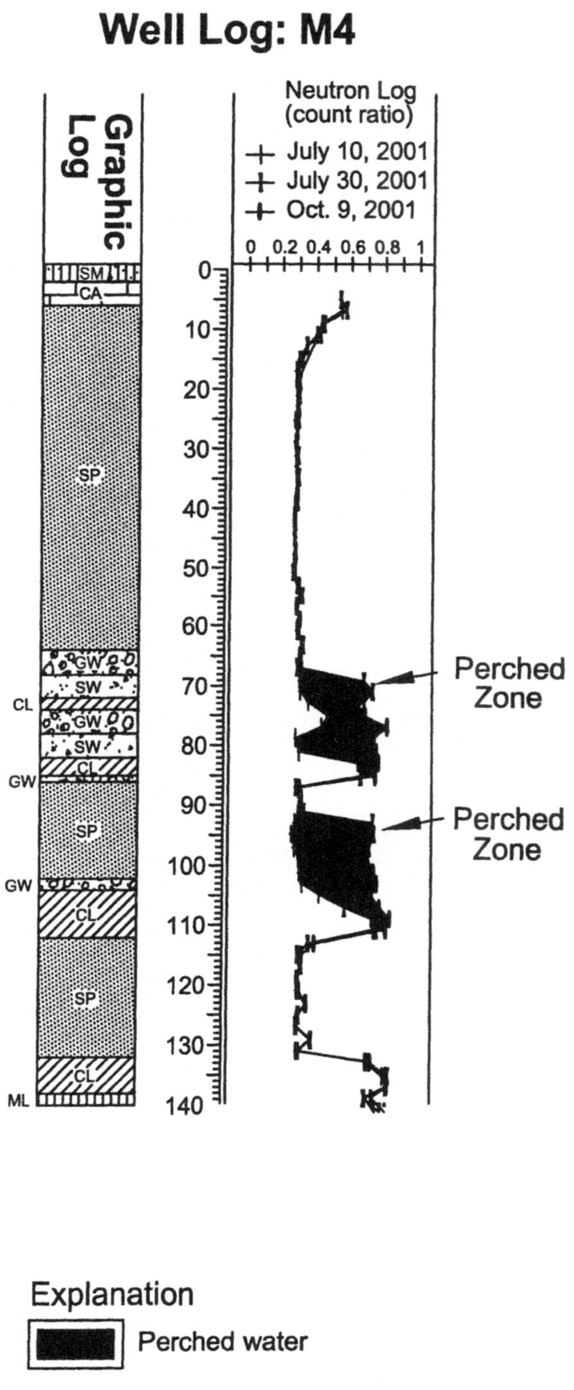

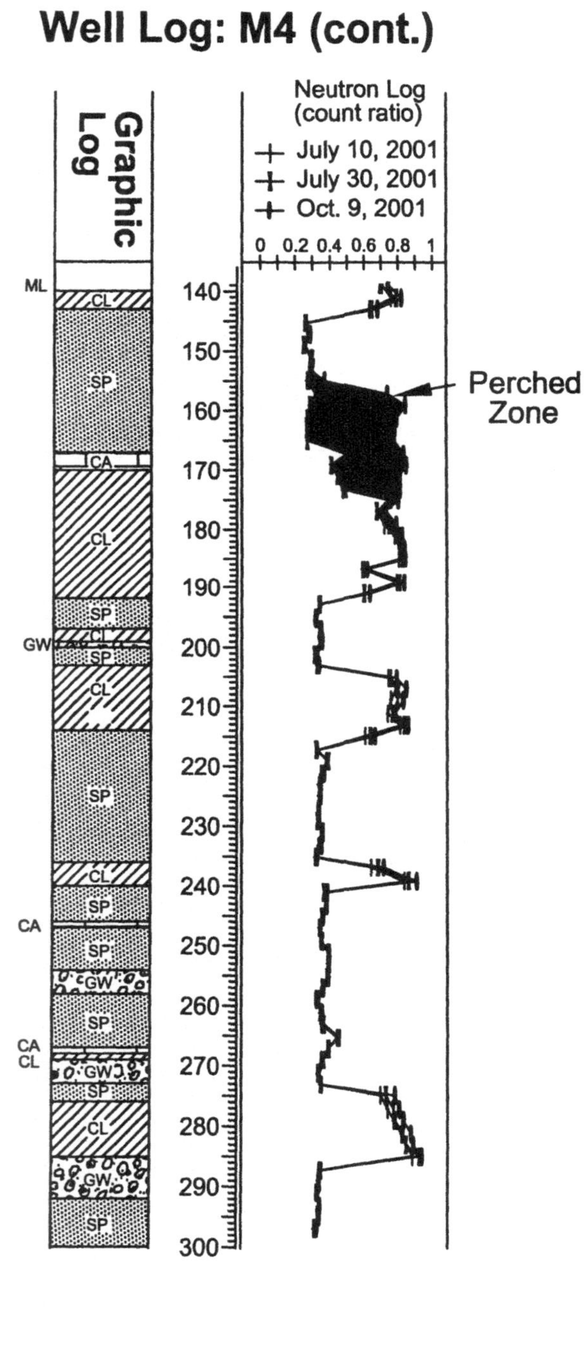

Figure 2.17 (continued) Neutron moisture logs showing perched water zones

CHAPTER 3
RESULTS AND DISCUSSION

ANALYSIS OF BASIN PERFORMANCE

For the entire data collection period, the basin recharged water at an average rate of 511 gpm, including no-flow periods, while water levels in the basin averaged about 0.5 ft over this same period. Cumulative flow into the basin from project start-up through July 2002 was approximately 283 million gallons (869 ac-ft), for an average inflow rate of 0.74 mgd during the total project reporting period of 384 days. Based on an average wetted footprint of approximately 0.4 ac, this corresponds to an average infiltration rate of about 1.8 mgd/ac or 6.5 ft/day of free water column. The average infiltration rate for the entire data collection period is conservative because flow into the basin prior to mid-December 2001 was underestimated. The flow meter and control system were not capable of recording flows in excess of 1,000 gpm during that period.

Flow-level and water-level data (presented in Appendix F) show little or no loss of infiltration capacity from start-up through mid-November 2001. During that period, the water level in the basin remained relatively constant at an average level of about 0.7 feet. Prior to mid-November, the average infiltration rate at the basin was approximately 7.9 ft/day of water column, based on totalized flow records for each month. In December, the water level in the basin began to increase more rapidly and cycle between the high and low settings due to a buildup of algae on the floor of the basin. The algae impeded the infiltration of water into the basin floor resulting in substantially lower infiltration rates, on the order of 2.2 ft/day of water column.

The basin floor was scarified and the layer of algae removed in January 2002 in an effort to improve the performance of the basin. Maintenance performed on the basin was effective in restoring the infiltration rate. After the basin floor was cleaned, the most accurate continuous periods of data collection were from January 14 through February 2, 2002 and from February 26 through March 20, 2002. During these two periods, the average infiltration rate was estimated to be 15.0 and 14.2 ft/day of water column, respectively (assuming a 0.4 ac wetted footprint), which is considerably higher than the average infiltration rate of 7.9 ft/day for the period prior to mid-

November. The increase in infiltration rate is largely a function of the increase in flow to the basin.

After March 20, there were no significant periods of continuous operation at the basin due to a combination of equipment failures and lack of available water from the FHWRP. Infiltration rates were not estimated for this period because of the sporadic operation of the basin and the limited amount of data collected.

An average porosity of 40 percent was assumed for the vadose zone sediments (primarily fine sand) based on laboratory test results for soil samples collected from test pits at the site. Based on the 386-foot vadose zone beneath the basin and an average porosity of 40 percent, the volume of air-filled pore space within the vadose zone beneath the 0.5 ac basin would be calculated as:

386 ft x 0.5 ac x 0.4 = 77 ac-ft of void space

Comparing this volume with the estimated 869 ac-ft of water that infiltrated during the year-long project, approximately 11 pore volumes have passed through the vadose zone to the water table. An estimate of the vertical rate of downward migration of the wetting front, or seepage velocity, can be obtained by dividing the average infiltration rate prior to mid-November (7.9 ft/day of water column) by the porosity (0.4) to yield a vertical seepage velocity of 19.8 ft/day.

This calculation assumes vertical seepage only, with no lateral spreading and no low permeability horizons that would cause perching of water. While this is not realistic, it nonetheless demonstrates that the large flux of water infiltrating into the floor of the basin would move more rapidly downward through the clean sand units until low permeability clays are encountered, at which point lateral spreading would occur.

Field monitoring results confirm the rapid downward migration of the infiltrating water beneath the infiltration basin. For example, piezometer M4/P3 completed atop a clay layer at 104 ft bgs first contained water on July 25, 2001, 15 days following basin start-up. On that date, approximately three feet of water had accumulated on that clay layer. Figure 2.16 shows the build-up of perched groundwater in M4/P3 after the water level datalogger was installed on July 30, 2001. Up to 11 feet of perched water was observed in piezometer M4/P3 beneath the infiltration basin. As shown in Figure 2.16, the perched water builds up quickly and dissipates quickly following an interruption in flow to the basin. The rate at which the perched water dissipates suggests that the perching clay beds are of limited lateral extent. Additional exploratory boreholes and monitor wells would be required to map the extent of the clay beds and the degree of lateral spreading of the wetting front.

The hydrograph for M4 (see Figure 2.15) located adjacent to the infiltration basin shows a 3-ft rise in water level during the period between August 7 and September 10, 2001. This rise, which began about 28 days following start-up, appears to signal the arrival of the wetting front at the water table. The arrival of water at the water table within one month of start-up is not unreasonable, given the high permeability of the sands underlying the basin. If the rise in water level is attributable to basin recharge, this suggests a vertical velocity for the wetting front of 13.8 ft/day (386 ft/28 days). This rate is lower than the predicted seepage velocity calculated from water application rates (19.8 ft/day). This difference is expected because the predicted rate assumes vertical seepage only with no lateral spreading and no low permeability clay or caliche layers. The rapid seepage velocities imply that the clay beds that are present in the vadose zone are not significantly retarding the downward migration of water toward the water table.

Appendix G shows shallow subsurface temperature profiles at each of the five thermocouple stations. Also shown are concurrent pool temperatures for stations A and D as well as air temperature measurements from the data logger located at the edge of the basin. Following are some preliminary observations based on these data, specifically the first 10 days after introduction of water into the basin.

- Station "A" located near the north edge of the basin (see Figure 2.11) was not yet submerged during the first week following start-up. Therefore, the surface thermocouple (0 cm depth) shows large diurnal temperature oscillations in response to heating and cooling of the surface of the sandy basin floor. Deeper thermocouples show significant attenuation of the amplitude of the diurnal temperature wave with depth in the relatively dry, unsaturated sand, which acts as a barrier to heat flow. The pool shoreline remained approximately midway between Stations "A" and "B" at the end of the period of record.

- Thermographs for Stations "B" and "C" located in the north central part of the basin (see Figure 2.11) indicate very rapid infiltration rates at these locations during the first 10-day period, as evidenced by the small attenuation and lag between the 30-, 60-, and 90-cm depths. The lag between the 30-, 60-, and 90-cm depths is so short that it appears as if there is only one line on the graph for the thermocouples at these depths. For example, at Station "C" the time lag for temperature maxima and minima between the 30- and 60-cm depths occurs only about 15 minutes apart, in which case it is not possible to discern between the lines on the graphs for these depths. This lag time suggests a vertical seepage velocity of approximately 1.2 m/hr or 94.5 ft/day. At Station "C" the amplitude of the thermal wave is virtually undiminished during travel through the first 90 cm of sandy sediments beneath the basin floor, which further confirms the rapid infiltration rate at this location.
- Thermographs for Stations "D" and "E" located in the southern part of the basin (see Figure 2.11) indicate somewhat slower infiltration rates than at Stations "B" and "C," as shown by the greater attenuation and lag of the diurnal temperature wave between the 30-, 60-, and 90-cm depths. For example, thermographs for Station "D" indicate a time lag of approximately 2 hours between the 30- and 60-cm depths, which translates to a wetting front movement of about 0.15 m/hr or 12 ft/day. At Station "D" the amplitude of the thermal wave is diminished by approximately half in the first 90 cm beneath the basin floor.

As summarized above, subsurface temperature profiles at each of the five thermocouple stations indicate very rapid seepage velocities. The observed rates of water movement using thermocouples bracketed the results calculated by dividing measured inflow by wetted basin area (19.8 ft/day). This result is expected because of the varying sediment permeabilities across the basin.

Neutron logging was performed prior to start-up to obtain background moisture profiles in the vadose zone. Background neutron logs (Figure 2.17) were run on February 1, 2001 (M1 and M3) and on July 10, 2001 (M4). As shown in Figure 2.17, background neutron logs show deflections (higher neutron counts) at each clay horizon, as clays generally have much higher water contents than do sands. Following start-up on July 10, 2001, subsequent neutron logs were run on July 12, July 30-31, and October 8-10, 2001 to monitor the progression of the wetting front through the vadose zone.

Neutron moisture logging confirms the rapid downward movement of water. By July 30-31, both of the monitor wells located adjacent to the infiltration basin (M3 and M4) showed the development of markedly wet zones at the 65 to 110 foot depths. This likely indicates saturated conditions due to buildup of perched water on the clay layer at 104 ft bgs (see M4 lithologic log, Figure 2.10). By the October 9, 2001 neutron logging event, a large perched water zone was evident from about 154 to 172 ft below surface in monitor well M4. Nearly 20 feet of perched water had accumulated on top of the clay layer at 170 ft bgs. These two major perching horizons were confirmed by the buildup of water in piezometers M4/P2 and M4/P3.

ANALYSIS OF DRY WELL PERFORMANCE

Graphs of hourly averaged water level and flow rate versus time for each month data were available are presented in Appendix F (Figure F.2). Gaps in the record indicate either that no water was available from the FHWRP, or that data was not recorded due to equipment problems.

The only accurate, relatively continuous period of data collection for the dry well was from August 24, 2001 through November 8, 2001. A total of 7.3 ac-ft of water (2.4 million gallons) flowed into the dry well, mostly prior to November 8, 2001. The sustained hourly average flow into the dry well was approximately 40 gpm at the beginning of September 2001, but declined to just under 20 gpm by the end of September, which was maintained until November 8, 2002 when the dry well was shut down due to equipment problems. The cause of the decline in infiltration capacity in the dry well is unknown, but is believed to be attributable to some form of well fouling or ineffective development of the well following its construction. Clogging of the borehole wall with clays generated during drilling is one possibility. While the drilling method employed is noted for being a relatively "clean" method of construction, it is likely that the method left behind residual clays, silts, and fine sands that are impeding movement of water from the well into the surrounding formation. It is possible that the clays swelled during the first two months of infiltration, with the result that permeability was significantly lowered. It is also possible that clays were smeared across areas of the formation during the extraction of the 20-inch drive casing. Biofouling of the wellscreen or filter pack is also a possibility, though this seems less likely given the relatively rapid decline in infiltration rates observed. Additional development of the well, including the introduction of surfactants, could not be accomplished within the limits of this project.

Poor performance of the well may also be related to air entrainment or the accumulation and/or trapping of air in the injection interval. The well was operated such that it cycled on and off frequently (every 30 minutes on average). As a result, there was a significant volume of air introduced to the well and formation. Operations of similar dry wells in the City of Scottsdale, Arizona showed that recharge capacity declined substantially as a result of air entrainment. In addition, when a moderate to high degree of air entrainment occurred there was no recovery of recharge capacity during a drying period. In other words, idle time did not reverse the effect of plugging due to air entrainment (Marsh, Small, and Close 1997).

The performance of the dry well was significantly below expectations in terms of injection rates. The quantity of water recharged and the limited operation of the dry well was insufficient to allow definitive conclusions related to infiltration rates and water quality impacts. The only conclusions made are based on the water level measurements recorded for M1, located adjacent to the dry well and the neutron logging results. The hydrograph for M1 (Figure 2.15) shows a 1.5-ft rise in water level during the period between September 4 and October 2, 2001. This rise, which begins about 55 days following start-up, may signal the arrival of the wetting front at the water table beneath the dry well. This implies a wetting front velocity, or seepage velocity, of about 4.2 ft/day (231 ft/55 days), which is substantially lower than the seepage velocity calculated for the infiltration basin. Vertical seepage velocity is largely a function of the recharge rate, which partially explains the lower seepage velocity calculated at the dry well. A subsequent 1.2-ft decline in the water level in M1 during the period between October 16 through October 25, 2001 may be attributable to the decline in flow rate into the dry well that occurred during September-October.

Neutron logging was performed prior to start-up to obtain a background moisture profile in the vadose zone beneath the dry well. A background neutron log (Figure 2.17) was run on July 10, 2001 for monitor well M1. As shown in Figure 2.17, the background neutron log shows deflections (higher neutron counts) at each clay horizon, as clays generally have much higher water contents than do sands. Following start-up on July 10, 2001, subsequent neutron logs were run on July 12, July 30-31, and October 8-10, 2001 to monitor the progression of the wetting front through the vadose zone. By the October 9, 2001 neutron logging event, large perched water zones were evident on top of the clay layers at 120 ft and 205 ft below surface. These two major perching horizons are confirmed by the increase in water level in M1.

ANALYSIS OF WATER QUALITY CHANGES

Samples for water quality analysis were collected by EPWU staff from the following locations:

- Fred Hervey Water Reclamation Plant effluent (source water to recharge facilities)
- Monitor well M1 (located 42 ft east of the vadose zone recharge well)
- Monitor well M4 (located 40 ft southeast of the infiltration basin)
- Piezometer M4/P2 (screened in vadose zone at 147-167 ft bgs)

Water quality results are summarized in Tables 2.4 and 2.5. A background groundwater sample was collected from monitor wells M1 and M4 several months prior to beginning recharge. Following the start of recharge in early July 2001, additional samples were collected periodically during the remainder of the project.

Background (pre-test) results demonstrate that the groundwater underlying the study area is of the sodium-chloride-bicarbonate type, with pH of approximately 7.6, and total dissolved solids (TDS) concentrations in the range of 440 to 560 mg/L. Nitrate-nitrogen (NO^3-N) concentrations are very low, approximately 1 mg/L.

The FHWRP product water has a slightly higher TDS concentration (620 mg/L), as compared to the background groundwater. The higher TDS appears largely attributable to slightly higher concentrations of sodium and chloride in the product water. In addition, nitrate-nitrogen concentrations are slightly higher than the ambient groundwater, averaging about 2.3 mg/L NO^3-N in the product water.

A perched water sample was collected from piezometer M4/P2 on October 25, 2001. On that date following the first three and a half months of infiltration, approximately 17 feet of perched water had built up beneath the infiltration basin on top of a 22-foot thick clay bed whose top lies at a depth of 170 ft below surface. The perched water sample had a TDS concentration of 680 mg/L, which is similar to that of the FHWRP product water. The observation that the TDS of the perched water is similar to that of the product water indicates that little soluble salts are present in the vadose zone beneath the infiltration basin. This finding is consistent with the clean sands observed during drilling of M4, as well as the results of laboratory EC tests performed on samples of the sand.

In spite of the somewhat higher TDS concentration of the FHWRP product water, major ion concentrations in groundwater samples collected from the two monitor wells have shown little change over the course of the project. This probably reflects the fact that some dilution of the product water with the underlying groundwater is occurring, given that the monitor wells are screened over approximately the upper 25 feet of the saturated zone. If infiltration were to continue over longer periods of time, it is expected that the underlying groundwater would more closely approach the chemical composition of the product water.

Turbidity values were elevated on some dates, probably as a result of the use of a bailer to collect the samples. For example, a very high turbidity was reported in the sample from M4 collected on April 2, 2002. The anomalously high concentrations of some metals (e.g. Al, Fe, Mn, Zn) are likely the result of acidification of the turbid sample, with the resulting dissolution of clay minerals. These elevated metals concentrations are not believed to be representative of true groundwater conditions.

Boron is frequently used as a conservative tracer of wastewater effluent, which generally contains elevated concentrations of this element. Boron concentrations reported for groundwater samples from M4 located adjacent to the infiltration basin (0.129 and 0.135 mg/L) appear similar to those reported for background groundwater samples from the Hueco Basin (Buszka et al., 1994). Product water from the FHWRP reportedly contains boron concentrations two to three times higher than this, or approximately 0.3 mg/L (Buszka et al., 1994).

RECHARGE FACILITY COST ANALYSIS

Cost data for the dry well, infiltration basin, and associated instrumentation and control systems was maintained by EPWU. Table 3.1 provides a summary of the cost of each facility, the associated instrumentation and control systems, and operations and maintenance. Instrumentation costs include costs for the pipeline from the reclaimed water pipeline to the recharge facility, the concrete valve vault, all valves, flow meters, the instrumentation shed and data acquisition equipment, and transducer. The operations and maintenance costs are for the one-year pilot study. The infiltration basin construction cost includes the cost to excavate the basin; construct the observation deck; and purchase and install the stilling well, staff gage, and perimeter fence. The dry well construction cost includes the cost to drill and construct the dry well and surface completion at the wellhead, which involved the connection of the 6-inch pipeline from the reclaimed water pipeline to the 4-inch recharge pipe, and installation of a pressure gage and air release valve. Instrumentation costs for the dry well are similar to the basin.

The total cost of the dry well and infiltration basin including construction, instrumentation, and operations and maintenance (O&M) was about $135,000 and $70,000, respectively. For the one-year pilot study the cost per unit of recharge capacity for the dry well is $4,500,000/mgd, based on a sustained recharge capacity of 20 gpm or 0.03 mgd. The cost per unit of recharge capacity for the basin is $61,000/mgd for the one-year pilot study, based on a sustained recharge capacity of 800 gpm or 1.15 mgd.

Table 3.1

Summary of recharge facility costs

Item description	Cost ($)	Sustained Recharge Capacity (mgd)	Cost per Unit of Recharge Capacity ($/mgd)
Dry well construction and surface completion	85,000		
Dry well instrumentation and O&M	50,000		
Dry Well Subtotal	135,000	0.03	$4,500,000
Basin construction	40,000		
Basin instrumentation and O&M	30,000		
Basin Subtotal	70,000	1.15	$61,000

OPERATIONAL DIFFICULTIES AND POTENTIAL IMPROVEMENTS

The infiltration basin and dry well were operated from July 2001 through July 2002. During that time there were several periods during which one or both facilities were off-line due to either a lack of supply to the facilities from the FHWRP, equipment problems, or maintenance. Each of these issues is discussed in more detail below.

The FHWRP was shut down on at least two occasions due to pump failures at that facility. Supply to the infiltration basin and dry well was discontinued until the pump was repaired. There were also several occasions starting in May 2002 and continuing through July 2002 when the demand for reclaimed water exceeded the supply. EPWU currently has several contracts to provide reclaimed water from the FHWRP. At times when demand for reclaimed water by these customers exceeded the supply, operations at both recharge facilities were discontinued.

There were several equipment problems that resulted in down time at both facilities. There were battery failures at both the basin and dry well. During these times, flow and water level measurements were not recorded. Batteries were initially selected as an alternate to AC power because of cost. The number and size of batteries selected was based on the predicted performance of the basin and dry well and the frequency at which the flow control valves would open and close. The electrical requirements for powering the dry well were significantly underestimated because the dry well cycled on/off much more than anticipated due to the poor performance of the well. In retrospect, the time required and labor cost associated with changing and recharging batteries every one to two weeks far exceeded the cost associated with running AC power to the site. The availability of AC power at the site would have greatly increased the amount of data collected.

Initially there was not sufficient grounding protection for the instrumentation and control system at the dry well and basin. Lightening strikes damaged the data acquisition equipment and transducers, shutting the basin and dry well down early in the data collection period. After additional grounding protection was added, this problem was eliminated.

The 4-inch PVC pipeline to the infiltration basin was sized based on the predicted performance of the basin and estimates of available water from the FHWRP. The total water supply available to the project was predicted to range from a low of 250 gpm to a maximum of

1,200 gpm. This supply was to be split among three recharge facilities (at this point an infiltration gallery was still being considered). The size of the supply pipeline to the basin was based on an estimated water supply to that facility of 200 gpm. Based on the expected flows to the basin, the flow meter installed initially was not capable of recording flows above 1,000 gpm. The availability of reclaimed water and the performance of the basin both significantly exceeded expectations. Because the flow meter was not adequately sized, flows to the basin were underestimated until the flow meter was replaced with a meter capable of measuring higher flows. An 8-inch pipeline to the recharge basin and meter capable of recording flows up to 3,000 gpm would have been more appropriate given the availability of reclaimed water and infiltration rates.

At the dry well it was not possible to reduce the flow to a rate at which the water level in the well remained constant. This was a function of the type of valves installed. In anticipation of flow rates in the range of 100 to 200 gpm, a manually operated gate valve was installed at the reclaimed water pipeline. However, the dry well was not able to maintain a recharge rate greater than 50 gpm without frequent cycling. A butterfly valve would have allowed for more progressive control of the flow at lower rates.

Initially the transducer in the dry well was set at 159 bgs, which is essentially the bottom of the well. There was a lot of turbulence at the bottom of the well due to the surge of water into the pipe every time the flow control valve opened. The flow rate into the well surged up to 300 gpm during this cycling. The surges in flow were reduced by resetting both the PRV and the gate valve, although only coarse adjustments were possible. The excessive turbulence at the bottom of the well damaged the transducer and caused the cable to become severely tangled. The 2-inch screened transducer pipe was replaced with vented 2-inch blank pipe. The transducer pipe was also strapped to the 4-inch recharge pipe for added stability. In addition, the transducer was raised by 10 ft to 149 bgs. This helped reduce the effects of turbulence on the transducer.

The infiltration basin was off-line for a three-week period for maintenance to remove a layer of algae that had built up on the basin floor. Intermittent cleaning of the basin floor is usually required with most infiltration basins. If the basin floor clogs with fine sediments or algae develops, a low-permeability layer (clogging layer) is created that impedes the downward flow of water. In this case, the basin may be more susceptible to the growth of algae because the recharge water is generally higher in nutrients and organic matter.

CHAPTER 4
SUMMARY AND CONCLUSIONS

PROJECT SUMMARY

The purpose of this research project was to compare and assess the feasibility and applicability of two methods of recharge, namely infiltration basins and dry wells. While recharge through deep injection wells is a proven technology in the northeast El Paso area, surface recharge through basins and dry wells had not been tested. This project generated information on the relative merits and deficiencies of these recharge methods with respect to construction, operations and maintenance, cost, performance, water quality impacts, and permitting requirements.

The project was accomplished in six tasks: (1) Site Selection, (2) Exploratory Test Pits and Infiltration Testing, (3) Exploratory Boreholes and Monitor Well Construction, (4) Design and Construction of Pilot Recharge Facilities, (5) Recharge Facility Operations, and (6) Monitoring Methods and Data Collection.

Under Task 1, a suitable site for investigation and testing of the pilot recharge facilities was selected. Three candidate sites were identified and compared on the basis of their proximity to a recharge water supply, proximity to existing water supply and recharge wells, favorable soils and surface geologic features, occurrence of known or suspected contaminant and/or brackish water plumes, prior history of recharge, persistence of caliche or impediments to infiltration, projected depth to water, and site access. The Well 41 site was selected for further investigation based on its performance with respect to the above listed criteria.

Under Task 2, exploratory test pit excavation was conducted to assess the variability of near-surface soil and hydraulic properties at the Well 41 site. Four test pits were excavated at the site. Detailed stratigraphic sections were measured and sediment samples collected for laboratory testing. At each test pit, a calcium carbonate cemented caliche layer was encountered from approximately two to five feet bgs. Falling-head percolation tests were performed to obtain a preliminary indication of potential infiltration rates at the site. Based on an evaluation of the percolation test results, ponded infiltration rates for a shallow recharge basin were estimated to be about two to three feet per day, and possibly higher (greater than 10 feet per day) if soil conditions at the site are more similar to Test Pit 4. The calculated mean saturated hydraulic conductivity beneath each test pit ranged from 2.8 ft/day to 18.0 ft/day. Soil EC test results indicated that the soluble salt content of soils near the surface ranged from 153 μS/cm to 2,935 μS/cm and varied considerably from one location to another. Higher salt concentrations corresponded with samples collected near the caliche unit. Locations for the pilot recharge facilities (infiltration basin and dry well) were selected based on favorable results of the exploratory testing at Test Pit 4.

Under Task 3, exploratory boreholes were drilled in an effort to evaluate the leaching potential of the vadose zone, and the effects of leaching on regional groundwater quality. It was estimated that the infiltration basin could mobilize salts during the recharge testing period and increase the TDS concentration at Well 41 by up to 20 percent. However, any increase in the TDS concentration at Well 41 and other nearby wells was expected to dissipate with time. Upon completion of exploratory borehole drilling, monitor wells were installed to allow measurement of soil moisture content, soil pore-water quality, and the rate of recharge advancement. Two deep monitor wells were constructed to approximately 400 ft, one adjacent to the dry well and one adjacent to the infiltration basin. An additional shallow monitor well was constructed to approximately 100 ft adjacent to the infiltration basin. Each monitor well was equipped with a neutron probe access tube for neutron moisture logging. In addition, up to three 1-inch diameter PVC piezometers were completed to impermeable layers, for the purpose of monitoring the accumulation of recharge water and to permit collection of water samples from perched water zones.

Under Task 4, pilot recharge facilities were designed and constructed. A 0.5-acre infiltration basin was constructed south of Well 41 and adjacent to a reclaimed water pipeline. The basin was excavated to a depth of about 8 to 10 feet, which was sufficient to remove a shallow caliche layer. The instrumentation and control system for the basin was configured to allow the water level in the basin to fluctuate between low and high control settings.

The dry well was completed above the water table in the unsaturated sediments of the vadose zone. The borehole was drilled using a cable tool rig to advance 20-inch steel casing to a depth of 172 ft. Upon completion of the borehole, 12.5-inch outside diameter steel casing was installed to a depth of 172 ft. Blank casing was installed from 0 to 92 ft and from 162 to 172 ft. The injection interval was screened from 92 to 162 ft. The well was equipped with a 4-inch recharge pipe and a 2-inch transducer pipe. A retrievable choke was installed in the 4-inch recharge pipe to create backpressure in the injection pipe and minimize the potential for air entrainment during start-up and operations. The instrumentation and control system for the dry well was identical to the infiltration basin.

Under Task 5, recharge facility operations were initiated on July 10, 2001 and continued for a period of approximately one year through July 2002. The source of recharge water for both the basin and dry well was tertiary-treated wastewater from the FHWRP.

In general, flows to the basin fluctuated from about 600 gpm to 1,500 gpm. The low and high settings for the water level in the basin were kept at one and two feet, respectively, for the duration of the data collection period. From start-up through mid-November, water levels in the basin remained relatively constant at about 0.7 ft. Water levels gradually increased through mid-December and began to fluctuate regularly between the high and low settings due to the gradual build up of a layer of algae on the basin floor. The basin was shut down to scarify the basin floor and remove the layer of algae that had accumulated at the surface. After recharge operations were resumed in mid-January, the performance of the basin improved significantly. Water levels remained below the high setting for the remainder of the data collection period.

Numerous operational difficulties were encountered at the dry well, including battery failures, lightening strikes which damaged the data acquisition equipment, and difficulties with the control system and transducer. As a result, the only accurate, relatively continuous period of data collection for the dry well was from August 24, 2001 through November 8, 2001. The sustained hourly average flow into the dry well was approximately 40 gpm at the beginning of September, and dropped to just under 20 gpm by the end of September. This rate was maintained until November 8 when the dry well was shut down due to equipment problems.

Task 6 involved the collection of data from July 2001 through July 2002. There were several periods during which the basin and dry well were off-line due to equipment problems, maintenance, and lack of available reclaim water supply from the FHWRP. When the basin and dry well were operational, flow and water level data were recorded every minute. Water levels were also continuously recorded for piezometer M4/P3 in M4 from 7/30/01 through 10/11/01 to monitor the build-up of perched groundwater at the top of the clay bed encountered at 170 ft bgs. Field water quality measurements and water quality samples and were collected periodically from M1, M4, and piezometer M4/P2.

Thermocouples were installed in the basin floor for temperature profiling in order to monitor the diurnal thermal pulses associated with recharge. Subsurface temperature profiling was used to estimate water recharge rates beneath the basin. Neutron logging was performed periodically to track the progression of recharge water en route to the water table.

Infiltration rates at the basin ranged from a low of 2.2 ft/day to a high of 15.0 ft/day of water column. The highest infiltration rates were observed at the start of the project and after maintenance was performed to remove the layer of algae that had accumulated on the basin floor. The lowest infiltration rates were recorded in mid-December when the layer of algae impeded the infiltration of water into the basin floor.

Field monitoring results confirm the rapid downward migration of infiltrating water beneath the basin floor. Water level measurements in piezometer M4/P3 adjacent to the basin demonstrate that water had begun to perch on the clay layer at 104 ft bgs within 15 days following recharge startup. Perched water in M4/P3 built up quickly and dissipated quickly following interruptions in flow to the basin. The rate at which the perched water dissipated is an indication that clay or caliche beds may be of limited lateral extent.

Water level measurements in M4 indicated the arrival of recharge water at the water table within one month following startup. This suggests a vertical seepage velocity of about 13.8 ft/day. This rapid seepage velocity indicates the clay beds present in the vadose zone do not significantly retard the downward migration of water to the water table.

Neutron logging results also confirm the rapid downward movement of water. By July 30-31, 2001 (20 days after start-up) neutron logging results for both monitor wells M3 and M4 show the development of wet zones at 65 and 110 ft bgs. By the October 9, 2001 neutron logging event (91 days after start-up), a large perched water zone was evident on top of the clay layer at 170 ft bgs in M4.

Thermocouple data proved to be a useful and inexpensive method of monitoring the variability of and changes in the permeability of the basin. Subsurface temperature profiles at each of the five thermocouple stations indicated very rapid infiltration rates. During basin startup, seepage velocities were estimated to range from 12.0 ft/day in the southern part of the basin to 94.5 ft/day near the north central part of the basin. These observed rates of water movement bracketed the measured seepage velocity of 13.8 ft/day calculated by dividing the depth to water by the number of days until arrival of recharge water at the water table.

Water quality results for samples collected from M1, M4 and piezometer M4/P2 show little change over the course of the project. This probably reflects the fact that significant dilution of the product water with the underlying groundwater is occurring. If infiltration were to continue over longer periods of time, it is expected that the underlying groundwater would more closely approach the chemical composition of the product water. The perched water sample from M4/P2 sample had a TDS concentration of 680 mg/L, which is similar to that of the FHWRP product water. The observation that the TDS of the perched water is similar to that of the product water indicates that little soluble salts are present in the vadose zone beneath the infiltration basin. This finding is consistent with the clean sands observed during drilling of M4, as well as the results of laboratory EC tests performed on samples of the sand.

The performance of the dry well was significantly below expectations in terms of injection rates. The quantity of water recharged and the limited operation of the dry well were generally insufficient to allow definitive conclusions related to infiltration rates and water quality impacts. The only accurate, relatively continuous period of data collection for the dry well was from August 24, 2001 through November 8, 2001. A total of 7.3 ac-ft of water (2.4 million gallons) flowed into the dry well, mostly during the period prior to November 8, 2001.

The poor performance and low infiltration capacity of the dry well may be attributable to some form of well fouling or ineffective development of the well following its construction. Clogging of the borehole wall with clays, silts, and fine sands that impede movement of water from the well into the surrounding formation is one possibility. Clays may have been smeared across areas of the formation during the extraction of the 20-inch casing. Biofouling of the wellscreen or filter pack could also reduce the infiltration capacity of the dry well. Poor performance of the well may also be related to air entrainment or the accumulation and/or trapping of air in the injection interval. The well was operated such that it cycled on and off frequently (every 30 minutes on average). As a result, there was a significant volume of air introduced to the well and formation.

COMPARISON OF RECHARGE METHODS AND APPLICATIONS POTENTIAL

This project generated useful information on the relative merits and deficiencies of an infiltration basin and dry well with respect to construction, operations and maintenance, cost, performance, and water quality impacts. A comparison of these recharge methods with respect to each of these factors is provided below.

The infiltration basin was relatively easy to construct compared to the dry well. The uppermost caliche layer encountered at the basin was relatively easy to remove with a large bucket excavator. The only problem encountered during construction was the instability of the basin walls due to the sands encountered below the caliche. This could have been avoided by benching or sloping the basin walls more. In comparison, the dry well was considerably more difficult to construct. Drilling contractors with the requisite qualifications, experience and licensing needed to construct the well as designed were extremely limited. The surface completion of the dry well was also more involved.

Operations and maintenance of the basin were less difficult than the dry well. The basin was in operation about twice as long as the dry well because equipment problems were less frequent. There were significantly more battery and transducer failures and problems with data acquisition at the dry well. The water level in the basin rarely cycled between the high and low setting, whereas, the dry well cycled on/off frequently. It was much more difficult to fine tune flow rates at the dry well because of the valves installed.

The total cost of the basin was substantially less than the dry well. The cost of the basin was on the order of $65,000 less than the dry well. In addition, the cost per unit of recharge was $61,000/mgd for the basin versus $4,500,000/mgd for the dry well.

The performance of the basin in terms of the quantity of water recharged far exceeded the performance of the dry well. However, the operation of the dry well and the quantity of water recharged was generally insufficient to allow definitive conclusions related to sustained recharge and infiltration rates. Pilot testing of dry wells in Scottsdale, Ariz. have shown dry wells capable of sustaining recharge rates up to 880 gpm (March, Small, and Close 1997). Therefore, the performance of the dry well constructed for this project should not be considered indicative of the performance of dry wells in general.

Water quality results for samples collected from monitor wells at both facilities generally show little change over the course of the project. Based on water quality data collected for this project, neither recharge facility appeared to have had a measurable impact on water quality. This probably reflects the fact that significant dilution of the product water with the underlying groundwater is occurring. If infiltration were to continue over longer periods of time, it is expected that the underlying groundwater would more closely approach the chemical composition of the product water. In addition, a perched water sample from M4/P2 had a TDS concentration similar to that of the FHWRP reclaim water. This is an indication that few soluble salts are present in the vadose zone beneath the infiltration basin.

In addition to comparing the two surface recharge methods, a comparison has been made of El Paso's existing injection wells and the infiltration basins evaluated in this study. The feasibility of using injection wells for storage and recovery of water is demonstrated by more than 15 years of operation of the recharge facility at the FHWRP. When comparing surface recharge facilities with El Paso's on-going recharge operations using deep injection wells, the following observations can be made. There are several advantages to the use of injection wells: evaporation losses are minimal; land requirements for facilities are modest; and the risks of contamination of recharge water by shallow, soluble deposits found within the unsaturated zone are minimized. The City's existing injection wells were designed to inject up to one million gallons per day on a near-continuous basis. The wells have performed satisfactorily, although they require frequent redevelopment and maintenance to sustain their injection capacity. The primary disadvantage is that injection wells are relatively expensive to construct. The cost for construction of an injection well, including instrumentation, controls, security fencing, a permanent pump for periodic surging and development, and a pump house is about $375,000. The cost per unit of recharge capacity for a single injection well is therefore about $375,000/mgd of recharge capacity, based on a sustained recharge capacity of about 1 mgd.

Surface recharge facilities can be cost effective compared to an injection well if adequate land is available and relatively inexpensive. The sustained recharge capacity of the infiltration basin is roughly 1.15 million gallons per day based on the results of this study, however, the real long-term capacity is likely much greater. The basin may have been capable of a maintaining a higher recharge rate, however, supply from the FHWRP was a limiting factor, in which case the maximum infiltration rate was not attained. The total capital cost of the infiltration basin was about $70,000. Because the basin was located on property owned by EPWU there were no costs associated with land purchase. The cost per unit of recharge capacity for the basin is $61,000/mgd of recharge capacity, based on a sustained recharge capacity of about 1.15 mgd.

Maintenance for the basin typically consists of intermittent earthwork for reshaping and stabilizing basin walls, cleaning of the pond bottom and removal of low-permeability fine sediments. Maintenance costs for the basin are estimated to be about $5,000 per year. Maintenance of an injection well is performed on an as-needed basis, averaging about once every five years. Maintenance activities include surging, swabbing, acidization, and mechanical brushing to remove plugging and blockage of the well screens. Maintenance costs for an injection well are estimated to be about $30,000 every five years, or about $6,000 in terms of an annual cost. Accordingly, maintenance costs for these recharge facilities are essentially the same.

In summary, infiltration basins appear to be a viable alternative to use of injection wells for recharge of the deep aquifer. Based on data collected in this study, they appear to perform well, are relatively inexpensive to construct, have modest maintenance costs, and are easy to operate. Caliche layers and clay beds in the vadose zone are not significantly retarding the downward migration of water toward the water table. In addition, water quality sampling results to date have not indicated any significant deterioration of recharge water or native groundwater.

CHAPTER 5
RECOMMENDATIONS TO UTILITIES

This report provides useful information to utilities on recharging a deep aquifer using infiltration basins and dry wells. Research completed during the project provides examples of methods used to characterize the vadose zone in areas being considered for recharge. In particular, this project advanced the understanding of geochemical changes to surface-recharged water and the rate of advancement of the recharge wetting front in arid environments. Two particular areas of concern with respect to the performance of a surface recharge facility were investigated which include:

- Low permeability geologic deposits, including caliche and clay layers, that could impede the vertical migration of recharge water en route to the water table; and
- The potential for degradation of recharge water quality due to the leaching of soluble minerals by the recharge water as it percolates through the vadose zone.

Based on data collected during this project, the caliche and clay beds appear to be of limited lateral extent and do not significantly retard the downward migration of water toward the water table. Neutron logging, thermocouple, and water level data all confirm the rapid movement of recharge water to the water table. Although the water quality data collected for this project is limited, the results for samples from monitor wells at both facilities generally show little change over the course of the project. There was no indication that the recharge water quality degraded due to leaching of soluble minerals. The TDS concentration of a perched water quality sample was similar to that of the reclaimed water which suggests that little soluble salts are present in the vadose zone beneath the infiltration basin.

The infiltration basin and dry well were compared on the basis of ease of construction, operations and maintenance, cost, performance, and water quality impacts. In general, the infiltration basin was less expensive and easier to construct, operate and maintain. Where land is available and relatively inexpensive, an infiltration basin would be an acceptable recharge method under conditions similar to those encountered in northeast El Paso. While the performance of the dry well constructed in this project is not considered indicative of dry wells in general, it demonstrates that performance is sensitive to well design, the drilling method selected, materials used, well development, and operation. Any one of these factors could contribute to poor performance of the well.

APPENDIX A
ELIMINATION OF AN INFILTRATION GALLERY FROM THE STUDY

TO: Chris Rayburn, Project Manager July 11, 2000
AWWARF Project No. 2529
Zhuping Sheng, EPWU

FROM: Research Team (Boyle Engineering Corporation/
Daniel B. Stephens and Associates)

SUBJECT: **Recommendations by EPWU to Eliminate Galleries From Test Facilities**

The Research Team offers the following comments on EPWU's recommendation to eliminate the gallery from the recharge testing program in light of their concerns over salt loading and subsequent contamination of the ground water environment in the vicinity of the recharge testing facilities. We believe the following factors should be considered in the decision to eliminate a facility:

1. Salt load estimates, as furnished in our recent report, suggest that salt loading will be greater for the demonstration basin than for the gallery.

2. In considering the impacts of full-scale facilities, it should be noted that most of the soluble salts lie below the probable constructed depth of both the basin and the gallery. Thus, a given water flux from either method will pass through a similar volume of the vadose zone. Therefore, in either the scaled up basins or galleries, the mass of salt flushed will be similar for a given volume of recharge.

3. There appears to be significantly more research and operational experience with recharge basins than for galleries. From a research standpoint, it may be beneficial to investigate the hydraulic and chemical performance of the gallery.

4. Galleries and wells can potentially be used in more developed areas such as parks or playing fields, while recharge basins require availability of large, undeveloped areas.

5. In considering permanent, full-scale recharge facilities, recharge basins will suffer evaporation losses, whereas the gallery will experience no such losses. However, evaporation losses are usually only a few percent of the water budget. Basins may require additional security measures to prevent access to or contamination of the surface ponds. The galleries, while still requiring restricted access, will not require the same level of security.

6. Vadose zone recharge wells have the most flexibility in deployment, but are (1) likely to cost the most on a unit basis; (2) are likely to require the cleanest water to operate; and (3) would bypass only about half of the salt in the vadose zone.

7. Many large municipalities in the southwest which operate recharge facilities appear to favor the use of basins, and appear to have managed the salt flushed from the vadose zone. If recharge basins also prove to be a viable option for El Paso, construction for a demonstration recharge basin would yield information which is directly applicable to a permanent, full-scale recharge project.

Each of the proposed methods has its strengths and weaknesses. For a municipality with complex and extensive long-term recharge needs, it is likely that effective site-specific use could be made for each of the methods. Basins are inexpensive to build and maintain but require more open area and potentially more security. Galleries can be used in more locations, but are probably more expensive to build than are basins on a unit recharge basis. The vadose zone recharge well is likely the most expensive option. It has the greatest flexibility in deployment because of its small footprint but probably requires the cleanest (and most expensive) water for operation.

The Research Team continues to encourage the construction and testing of all three facilities (dry well, infiltration basin, and gallery). From a salt loading perspective, the preliminary calculations we provided indicate a relatively low risk of contamination to a level where drinking water standards are exceeded at Well 41, and are generally consistent with expectations about salt loading at the time of the project's inception. At the same time, we understand EPWU's desire to protect the ground water environment from contamination by salts which may be leached in the course of facility operations.

If it is determined to eliminate the gallery from the research program, the locations of the remaining facilities can remain as designed. If it is determined to eliminate the basin from the research program, EPWU may want to consider relocating the gallery to the site originally planned for the basin. This change will allow for a much shallower excavation and construction depth for the gallery.

Any such changes which are being considered should be evaluated in terms of their impact to layout and sizing of the various pipelines, valves, meters, and instrumentation and control systems.

APPENDIX B
LIST OF PROJECT RESOURCE MATERIALS

Table B.1
Project Bibliography

#	TITLE	AUTHOR	DATE	PUBLISHER
1	Evaluation of Ground-Water Resources in El Paso County, Texas	Ashworth, John B.	1990	Texas Water Development Board
2	Summary of Hydrologic Information in the El Paso, Texas Area with Emphasis on Ground-Water Studies, 1903 - 1980	White, Donald E.	1983	U.S. Geological Survey
3	Selected Hydrologic Data from the Northern Part of the Hueco Bolson, New Mexico and Texas	Orr, Brennon R., and Robert R. White	1985	U.S. Geological Survey
4	Evaluation of Ground-Water Resources in El Paso County, Texas	Ashworth, John B.	1989	Texas Water Development Board
5	Digital Model for Simulated Effects of Ground-Water Pumping in the Hueco Bolson, El Paso Area, Texas, New Mexico and Mexico	Meyer, W. R.	1976	U.S. Geological Survey
6	Ground-Water Resources of the Hueco Bolson, Northeast of El Paso, Texas	Knowles, Doyle B. and Richard A. Kennedy	1956	U.S. Geological Survey and Texas Board of Water Engineers
7	Ground-Water Development in the El Paso Region, Texas with Emphasis on the Lower El Paso Valley	Alvarez, Henry J. and A. Wayne Buckner	1980	Texas Department of Water Resources
8	A Preliminary Assessment of Land-Surface Subsidence in the El Paso Area, Texas	Land, L.F., and C.A. Armstrong	1985	U.S. Geological Survey
9	Computer Index of Public Service Board Geophysical Logs	El Paso Public Service Board	1989	El Paso Public Service Board

Table B.1
Project Bibliography

#	TITLE	AUTHOR	DATE	PUBLISHER
10	Ground-Water Resources of the El Paso Area, Texas	Sayre, A. N. and Penn Livingston	1945	U.S. Geological Survey
11	Geologic And Hydrologic Studies Of Sites HU1A And HU1B In Hudspeth County, Texas	Kreitler,C.W., J.A.Raney, W.F.Mullican III, E.W.Collins, and R.Nativ	1987	Bureau of Economic Geology, Univ. of Texas, Austin
12	Soil Survey : El Paso County, Texas	Jaco,Hubert B., Soil Conservation Service	1971	Soil Conservation Service, U.S. Dept. of Agriculture
13	Simulated Effects of Ground Water Pumping in Portions of the Hueco Bolson in Texas and Mexico during the period 1973 through 2029	Knowles,Tommy R. and Henry J. Alvarez	1979	Texas Department of Water Resources
14	Geohydrology and Potential Effects of Development of Freshwater Resources in the Northern Part of the Hueco Bolson, Dona Ana and Otero Counties, New Mexico	Orr, Brennon R. and Dennis W. Risser	1992	U.S. Geological Survey
15	Summary of Data From the 1981-83 Pilot Study and 1985-1989 Operations of the Hueco Bolson Recharge Project, Northeast El Paso, Texas	White, D.E. and Gail J. Sladek	1990	U.S. Geological Survey
16	Appraisal of Potential for Injection-well Recharge of the Hueco Bolson with Treated Sewage Effluent -- Preliminary Study of the Northeast El Paso Area	Garza, Sergio, Edwin P. Weeks, and Donald E. White	1980	U.S. Geological Survey

Table B.1
Project Bibliography

#	TITLE	AUTHOR	DATE	PUBLISHER
17	Aquifer Storage and Recovery Investigation	Boyle Engineering Corporation and Parsons Engineering Science, Inc.	1995	New Mexico/Texas Water Commission
18	Aquifer Storage and Recovery Final Appraisal Report, Supplemental Investigations of Recharge Basins	Boyle Engineering Corporation and Engineering Science, Inc.	1997	New Mexico/Texas Water Commission
19	Research Memorandum No. 9 - Injection Well Analysis	Parkhill, Smith & Cooper, Inc.	1992	El Paso Water Utilities Public Service Board
20	Proposed Groundwater Recharge, Chapter XVI	Parkhill, Smith & Cooper, Inc.	undated	El Paso Water Utilities Public Service Board
21	Preliminary Report on the Hueco Basin Recharge Project and the Treatment Process Required	Parkhill, Smith & Cooper, Inc.	1978	El Paso Water Utilities Public Service Board
22	Hydrogeology and Selected Water-Quality Aspects of the Hueco Bolson Aquifer at the Hueco Bolson Recharge Project Area, El Paso, Texas	Buszka, Paul M., Robert D. Brock, and Richard P. Hooper	1994	U.S. Geological Survey
23	Hydrogeologic and Water-Quality Data from Wells near the Hueco Bolson Recharge Project Area, El Paso, Texas, 1990 and 1991	Brock, Robert D., Paul M. Buszka, and Edward M. Godsy	1994	U.S. Geological Survey
24	Hydrology of the Shallow Aquifer and Uppermost Semiconfined Aquifer Near El Paso, Texas	White, Donald E., E.T. Baker, Jr., and Roger Sperka	1997	U.S. Geological Survey

Table B.1
Project Bibliography

#	TITLE	AUTHOR	DATE	PUBLISHER
25	Recharge Well Technology at the Water Campus Project, Scottsdale, Arizona	Marsh, Floyd L., Leonard L. Dueker, and Gary G. Small	undated	
26	Vadose Zone Recharge Wells: Field Evaluation of an Innovative Design	Marsh, Floyd L., Gary G. Small, and Christine Close	1997	Paper Presented at the 8th Biennial Sumposium on the Artificial Recharge of Groundwater, Tempe, Arizona
27	Vadose Zone Recharge Well Pilot Testing: An Innovative Option for ASR	Marsh, Floyd L., Gary G. Small, and Christine Close	undated	
28	Groundwater Occurrence of the El Paso Area and Its Related Geology	Tom Cliett	1969	New Mexico Geological Survey 20th Field Conference Guidebook, The Border Region
29	Geohydrologic Site Characterization of the Municipal Solid Waste Landfill Facility, U.S. Army Air Defense Artillery Center and Fort Bliss, El Paso County, Texas	Abeyta, Cynthia G.	1996	U.S. Geological Survey, Water-Resources Investigations Report 95-4217

APPENDIX C
TEST PIT EXCAVATION DATA

Table C.1

Test pit laboratory results

Sample	Initial Moisture Content Gravimetric (%, g/g)	Initial Moisture Content Volumetric (%, cm^3/cm^3)	Dry Bulk Density (g/cm^3)	Wet Bulk Density (g/cm^3)	Calculated Porosity (%)	Hydraulic Conductivity (cm/sec)	Particle Size Characteristics (mm)	(mm)	(mm)	C_u	C_c
TP-1	12.8	17.8	1.40	1.57	47.3	2.4E-04	0.015	0.26	0.41	27	1.4
TP-2	13.4	20.9	1.56	1.77	41.1	9.0E-05	0.021	0.24	0.36	17	1.6
TP-3	13.0	20.2	1.56	1.76	41.2	2.5E-05	0.0035	0.16	0.27	77	3.9
TP-4	12.9	19.7	1.52	1.72	42.5	1.0E-03	0.12	0.24	0.28	2.3	0.86

Notes:

d_{50} = Median particle diameter

Reported values for d10, Cu, Cc, and soil classification are estimates, since extrapolation was required to obtain the d10 diameter

APPENDIX D
ELECTRICAL CONDUCTIVITY TEST RESULTS

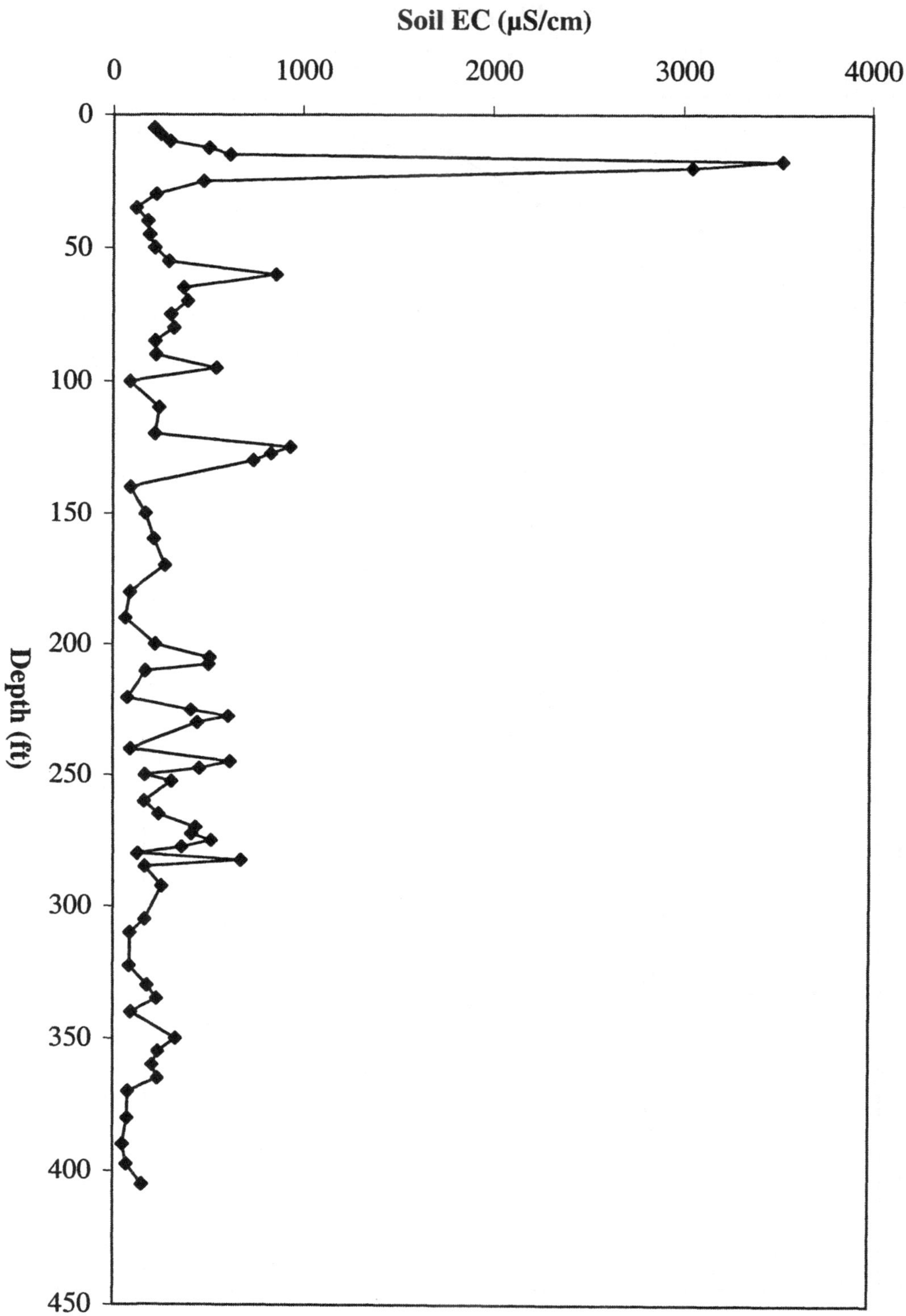

Figure D.1 Boring M1 Soil EC Profile

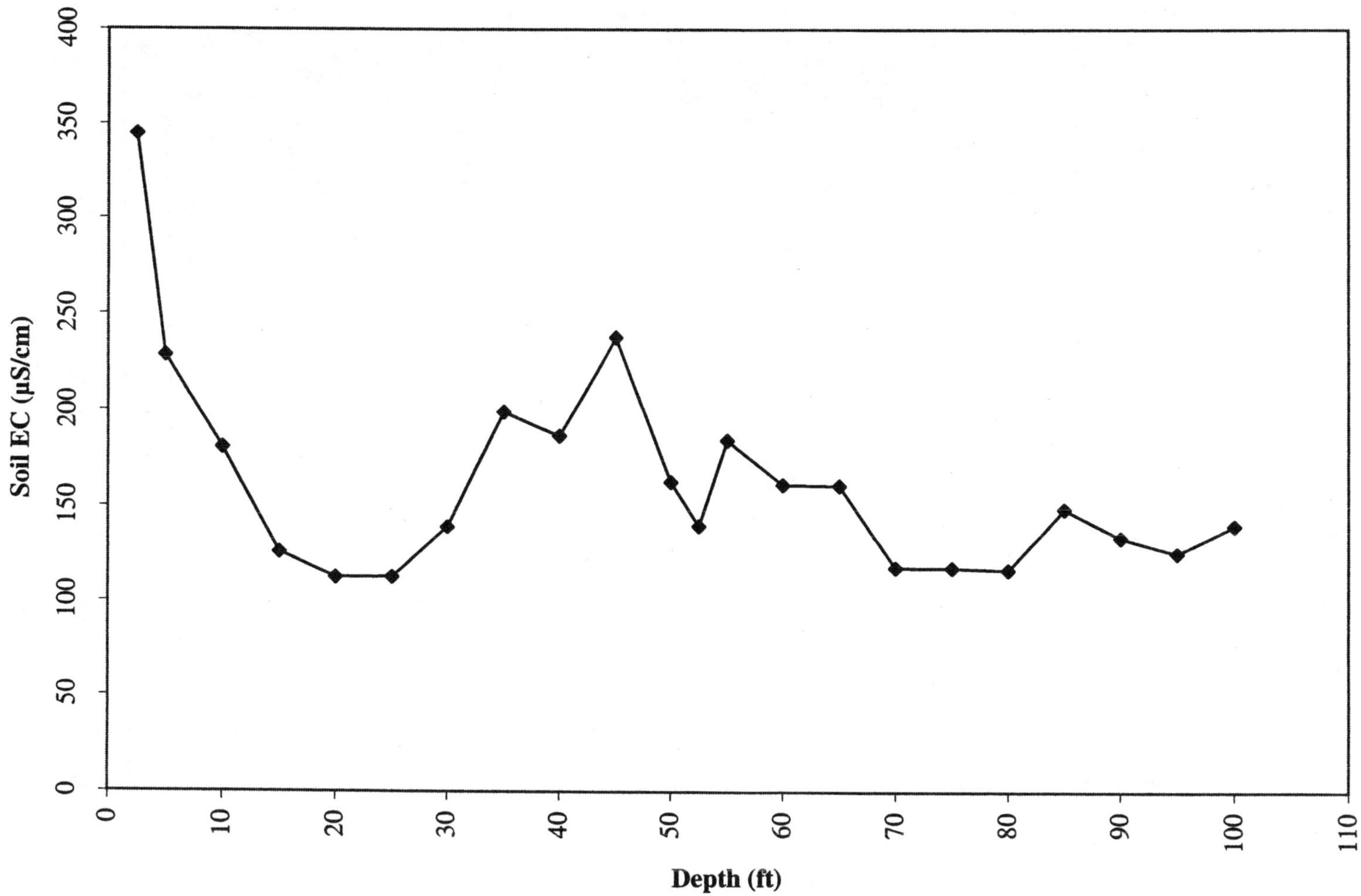

Figure D.2 Boring M3 Soil EC Profile

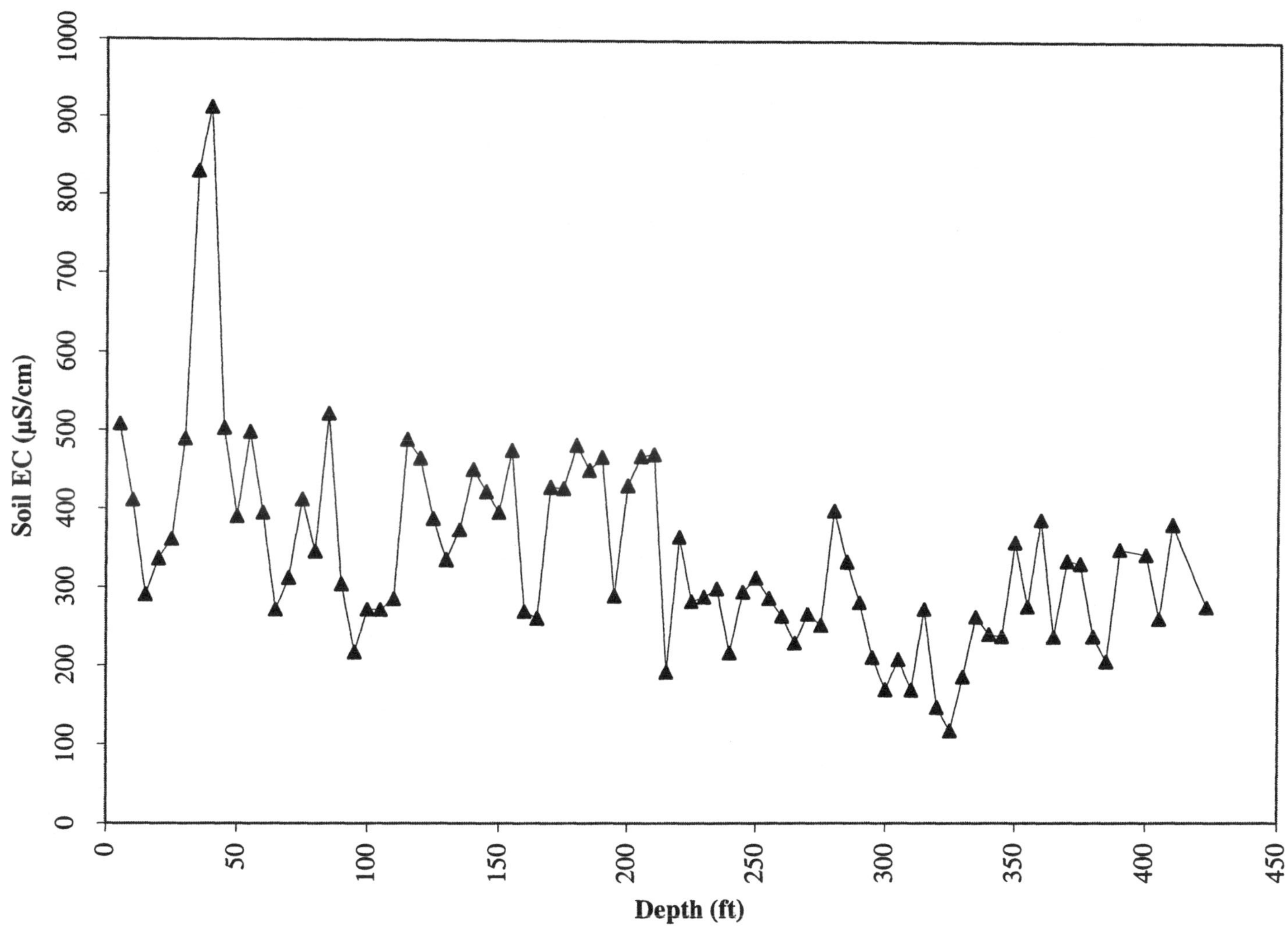

Figure D.3 Boring M4 Soil EC Profile

APPENDIX E
SALT LOADING ANALYSIS

AWWARF Project No. 2529
Preliminary Estimates of Salt Loading by Test Recharge Facilities

This paper presents preliminary estimates of the potential for mobilizing of salts contained within the unsaturated soil column, and estimates of the potential water quality changes which may result from operation of the test recharge facilities. The hydrologic and geochemical behavior of this system is highly complex. The analyses which follow are based on a number of simplifying assumptions relative to this highly complex system. As a result, the findings contain a high degree of uncertainty. The results are considered to provide order-of-magnitude estimates only. A more rigorous analysis of the behavior of this system, if required, could be accomplished through numerical modeling of the flow and solute transport processes.

Soluble salts that could be mobilized by infiltrating water were estimated by measurement of electrical conductivity on 1:1 soil:water extracts. The average soil bulk density in the field was assumed to be 1.6 g/cm3 and the average volumetric water content was assumed to be 5% for the following calculations and graphs.

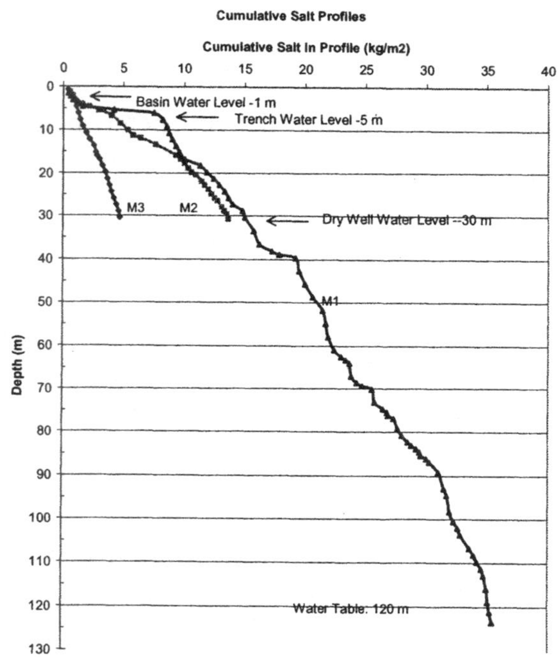

Cumulative salt profiles are shown in Figure 1 at the right. Samples were collected at approximately 2.5 foot intervals at each of the three boring locations on all three wells drilled. M1 and M2 show similar profiles while M3 (the proposed location of the infiltration basin) shows lower salt concentrations than the first two borings. Most of the variation among the three profiles is in the top 15 meters. No information is available on variation in salt mass below 30 meters because only one boring was drilled deeper than this depth (M1).

We have made simple estimates of salt loading in a scaled up infiltration system, salt loading at the pilot demonstration scale, peak salt concentrations that may be seen in a piezometer, and the increase in salt concentration that might be observed at Well 41 from a recharge facility. For the following calculations, data from boring M1 were used because it had the highest salt concentrations measured, and because soil samples were available for the entire thickness of the vadose zone.

Salt flushed by one-dimensional piston displacement. This calculation is considered to represent the scale up of a test demonstration to a large area where all vadose zone salt in an affected interval is anticipated to be eventually flushed.

The proposed square test infiltration basin is approximately 150 feet (46 meters) on a side for an area of 2,116 m2 (0.5 acre). From figure 1 (M1 data), approximately 74,060 kg of salt would be anticipated to be flushed (2,116 m2 * 35 kg/m2). This estimate represents a "worst case" scenario.

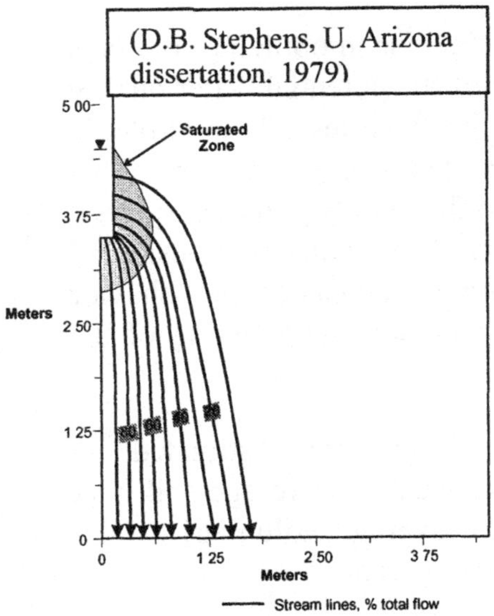

The infiltration gallery is approximately 100 feet (30 meters) long and two feet (0.6 meters) wide for an area of 18 square meters. Water in the trench would be ponded to approximately 5 meters below grade. For preliminary calculations, flow lines are estimated to extend approximately five meters laterally on either side of the trench (see figure 2 at the left), for a total footprint wetted area of 318 m2. The salt in the profile from 5 meters to the water table is approximately 30 kg/m2. Thus, for this small footprint with only five meters of lateral flow, one-dimensional flow would flush 9,540 kg of salt (10.6m * 30m *30 kg/m2).

The dry well footprint for 'one-dimensional' flushing is calculated with similar assumptions as for the infiltration gallery. The dry well is approximately one meter (36 inches) in diameter and radial flow lines are assumed to extend out five meters from the well bore for a footprint of 95 square meters ((11/2)2)*3.14). M1 shows that salt concentrations are higher in the top 30 meters than in the 100 meters below. The top 30 meters of M1 has 15 kg of salt/m2 while the next 100 meters has an additional 20 kg of salt/m2. Flushed salt would be about 1900 kg (95 m2 * 20 kg/m2). Thus, a dry well with top of screen at 30 meters depth will result in flushing nearly half of the salt present in the entire vadose zone profile.

Salt flushed by demonstration-scale testing. Unlike larger scale facilities, small scale testing results in significant lateral flow in a layered alluvial system. This increases the volume of soil flushed and the mass of salt mobilized. Preferential flow that is also likely to occur will tend to reduce the volume flushed. Overall, however, more salt can be expected to be mobilized in a small scale demonstration. For conservative calculations, we have simply added an order of magnitude to the salt mobilized for the two smaller demonstrations. Thus, conservative estimates are for 95,000 kg for the infiltration gallery and 19,000 kg for the dry well. For the larger area and smaller perimeter of the infiltration basin, we have applied a volume factor of three to obtain approximately 250,000 kg of salt flushed beneath the half acre test basin.

Salt concentrations anticipated at the water table. Based on the soil EC data, the average TDS of vadose zone water in M1 is estimated to be approximately 5,800 mg/liter. If this water undergoes simple piston displacement, then this would be the peak groundwater concentration likely to be encountered at the water table immediately beneath the test facilities.

The salt concentration which may be anticipated at the water table may be estimated using an alternative approach, as follows. Taking the infiltration basin as an example, it has been estimated that approximately 250,000 kg of salt may be flushed. Assume this mass will be flushed by the volume of water infiltrated by the basin over a one-year period, taken to be the design hydraulic loading rate (200 gpm) operating about 60 percent of the time for a period of one year. The total volume of water available to carry this mass is therefore approximately 240 million liters. This yields a theoretical added concentration of 250 million grams/240 million liters or about 1,000 milligrams per liter. It should be noted that this assumes uniform mixing of the salt mass in the recharge water and therefore represents an *average* added concentration. Peak concentrations may be significantly higher than this.

Salt concentrations anticipated at Well 41. The well is assumed to draw water radially. The recharge facilities are 300 meters (basin), 600 meters (gallery), and 730 meters (dry well) away from the Well 41, and have an average diameter of about 46 meters for the recharge basin, 11 meters for the dry well and 18 meters for the infiltration gallery. If we assume relatively little mounding, we can estimate dilution by comparing the average diameter of the recharge facility with the proportion of the Well 41 capture zone that it occupies.

As a simple approximation, the infiltration basin occupies an area equivalent to a wedge accounting for approximately 2.5% of the total radial flow field surrounding Well 41. The other recharge facilities (gallery and dry well) occupy a negligible portion of the flow field as a result of their greater distance from Well 41 and/or their smaller footprint. Calculated pretest TDS concentrations for groundwater from Well 41 are approximately 650 mg/L (EPWU 1999 data). Assuming piston displacement of the poor quality vadose zone pore water with TDS = 5,800 mg/L beneath the 0.5 acre pilot test basin, we estimate the maximum TDS that could result at Well 41 using a weighted average approach as:

$$
\begin{aligned}
TDSmax &= 0.975\ (650\ mg/L) + 0.025\ (5{,}800\ mg/L) \\
&= 778\ mg/L
\end{aligned}
$$

The hypothetical TDS increase at Well 41 from 650 to 778 mg/L represents a 20% increase. Observed increases in TDS are expected to be lower as a result of the conservative nature of the assumptions used in the calculations described above. In addition, any increase in TDS concentration at Well 41 and other nearby wells is expected to be transient, and should dissipate with time.

APPENDIX F
WATER LEVEL AND FLOW DATA

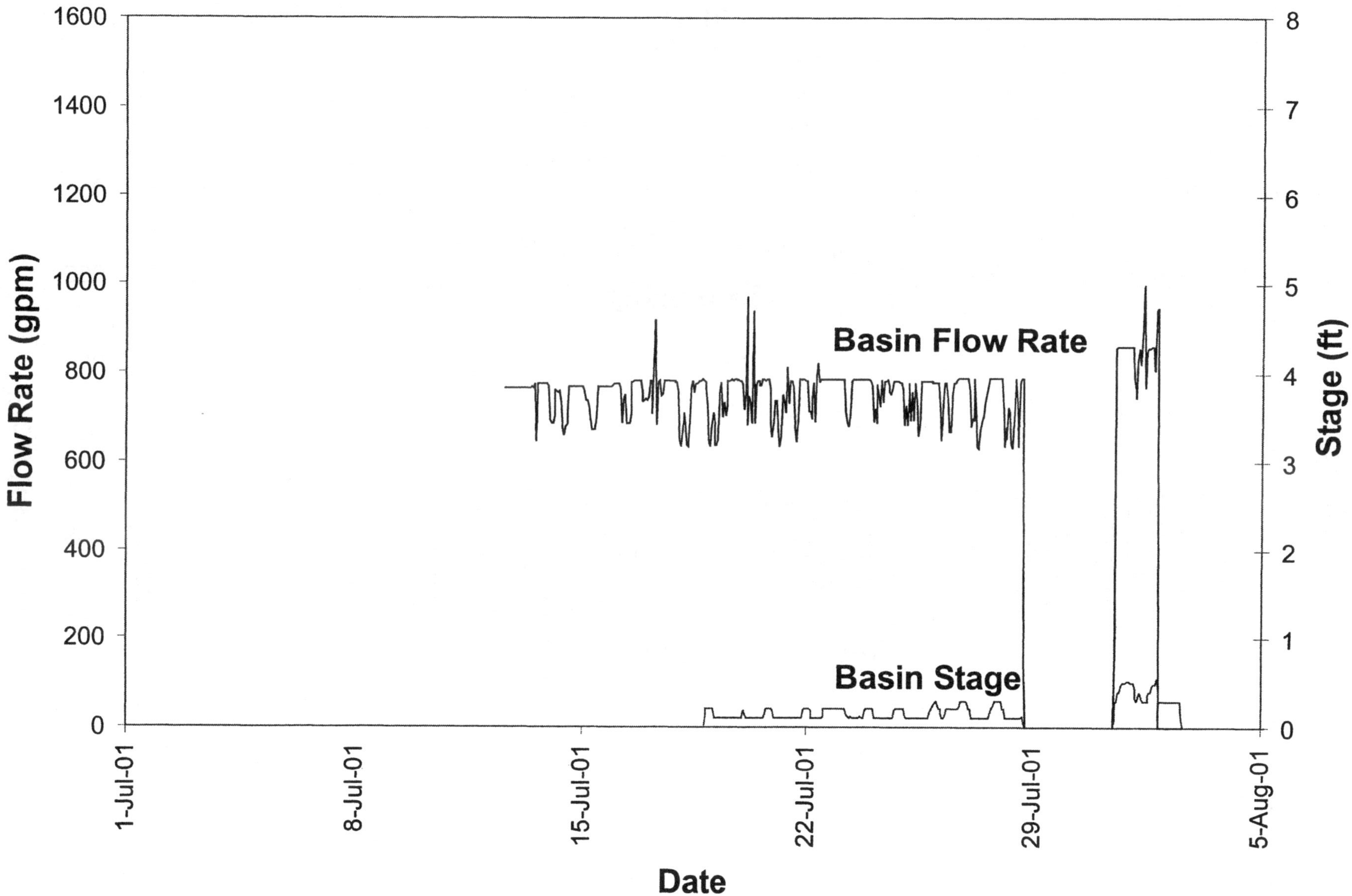

Figure F.1 Basin Flow Rate and Stage Versus Time, July 2001

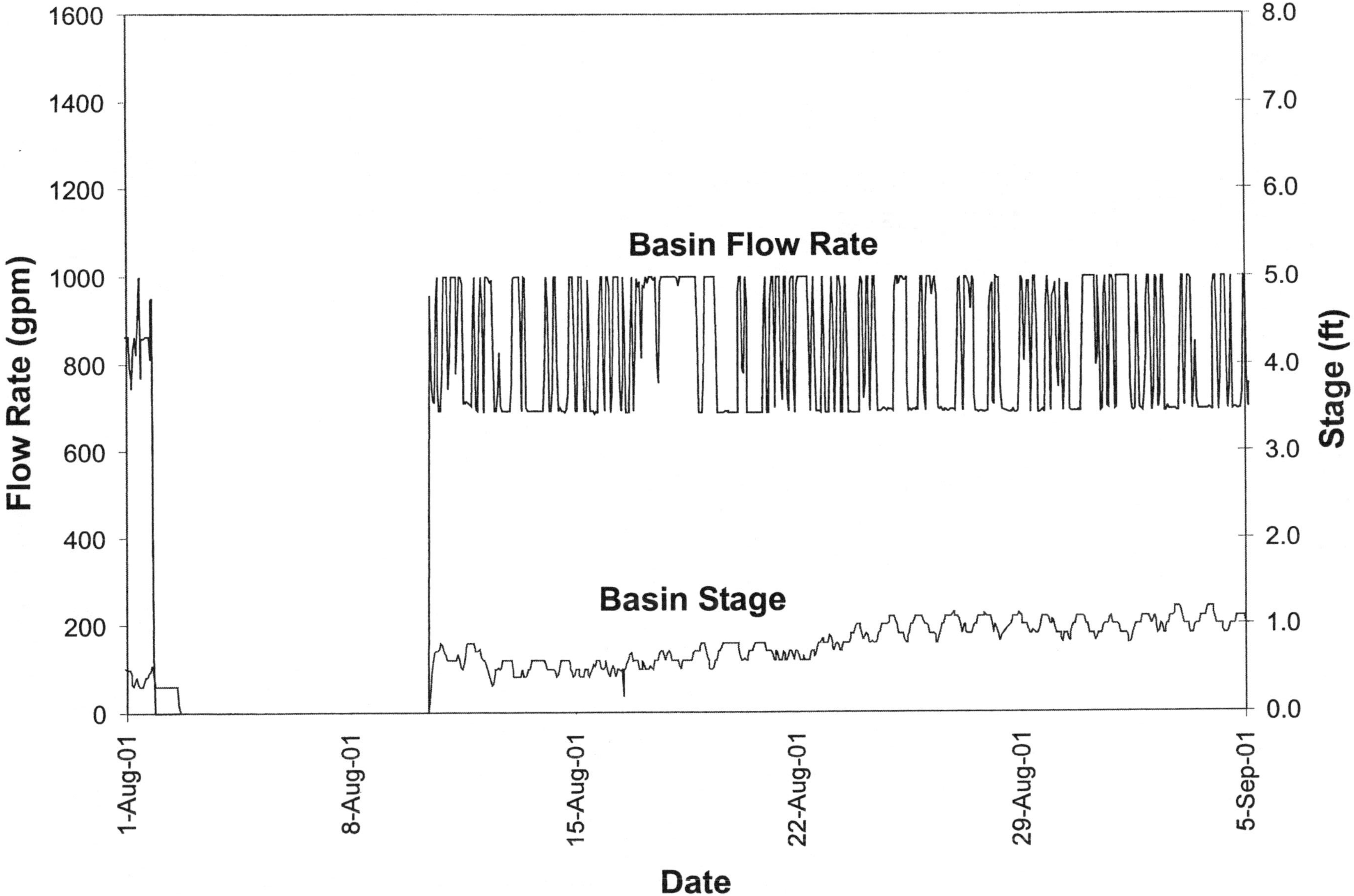

Figure F.1 (continued) Basin Flow Rate and Stage Versus Time, August 2001

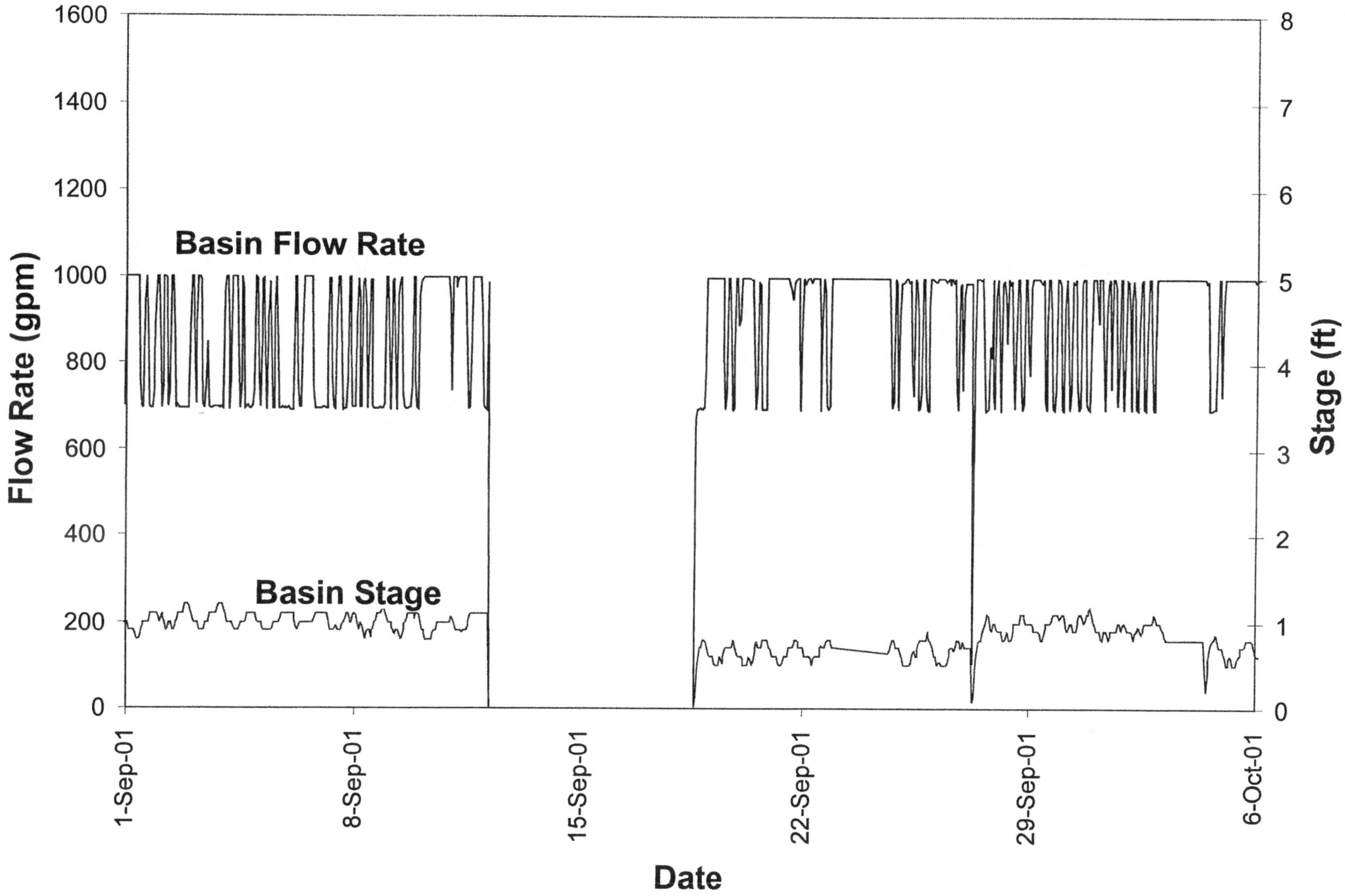

Figure F.1 (continued) Basin Flow Rate and Stage Versus Time, September 2001

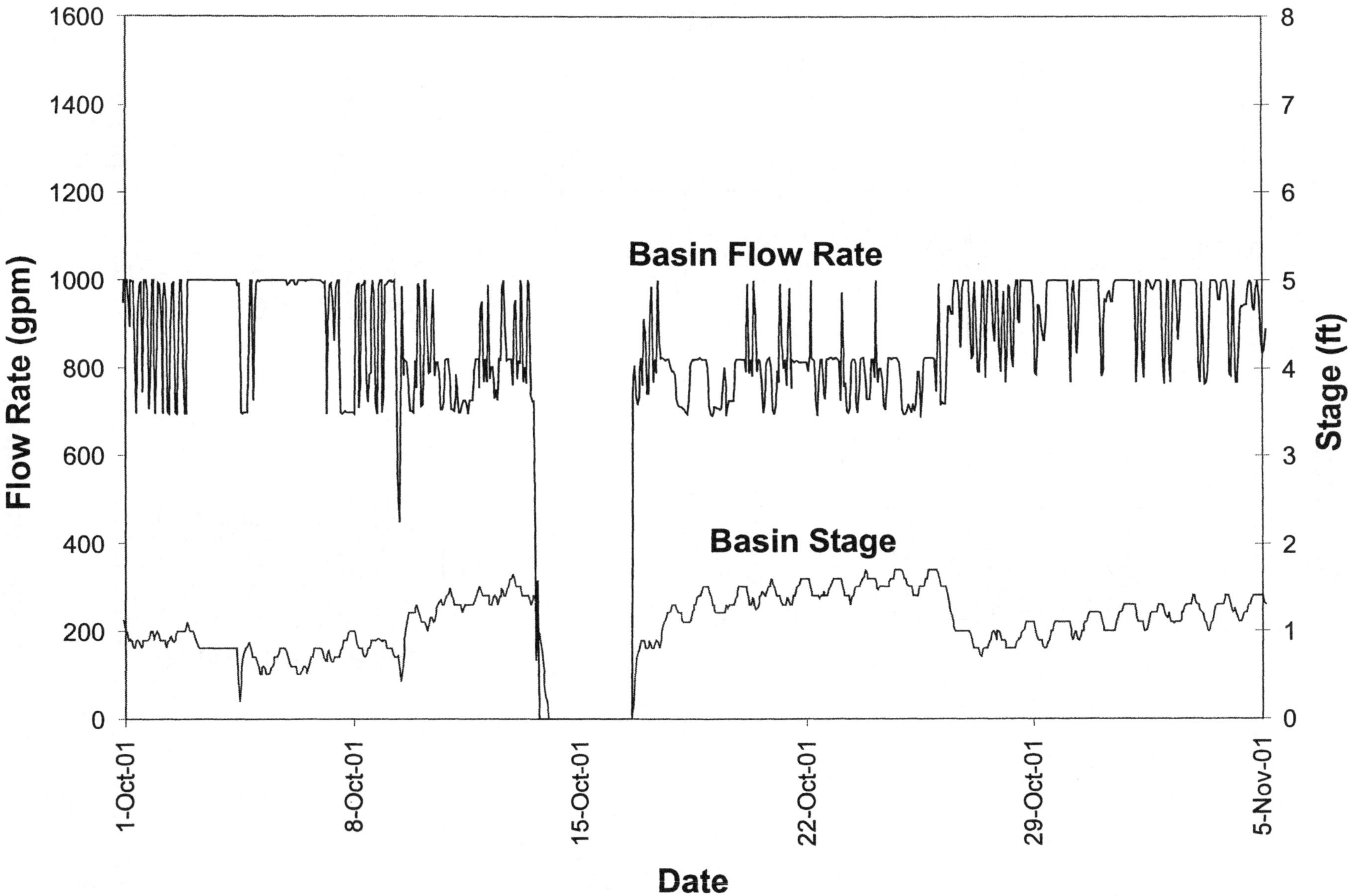

Figure F.1 (continued) Basin Flow Rate and Stage Versus Time, October 2001

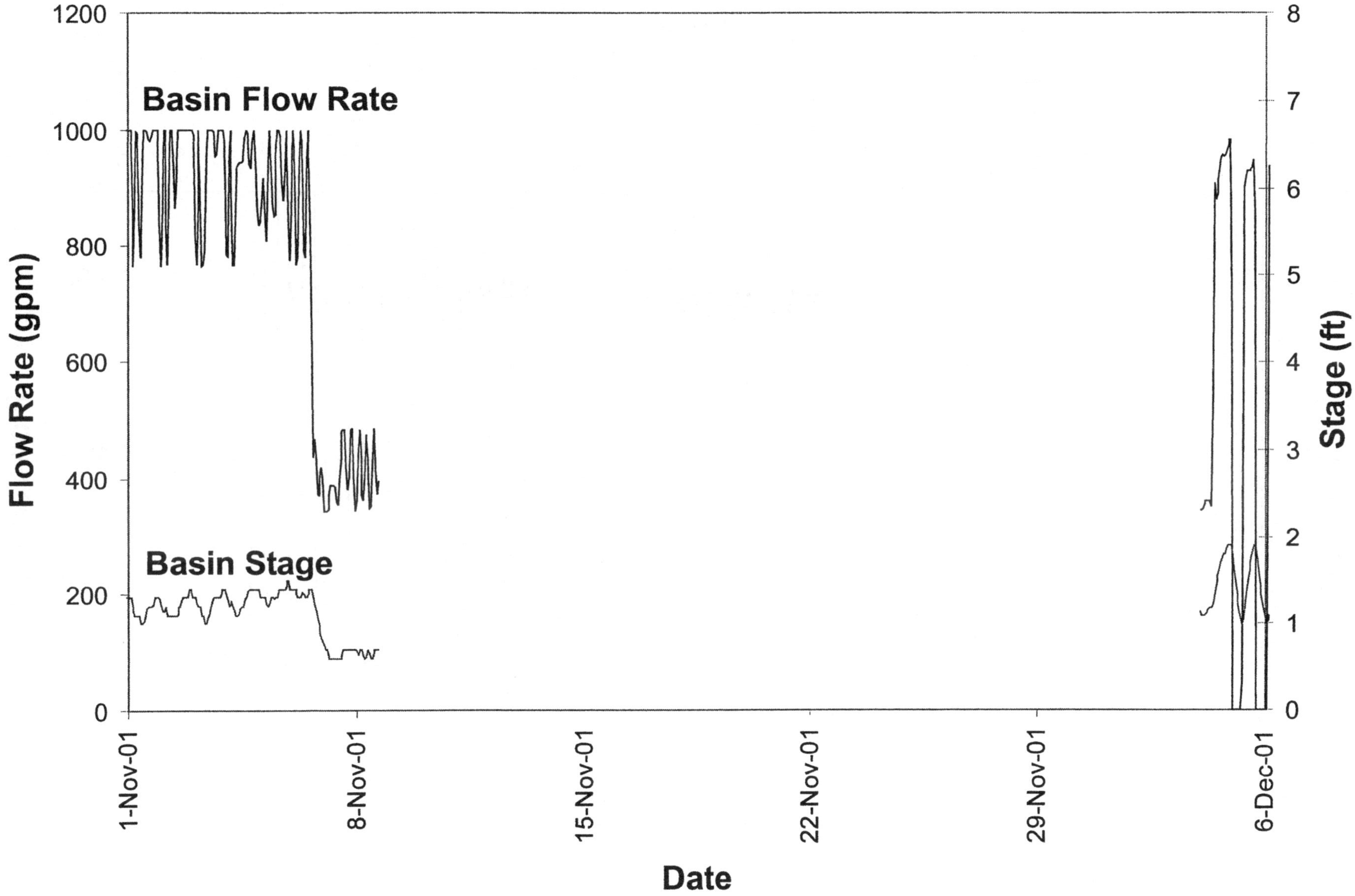

Figure F.1 (continued) Basin Flow Rate and Stage Versus Time, November 2001

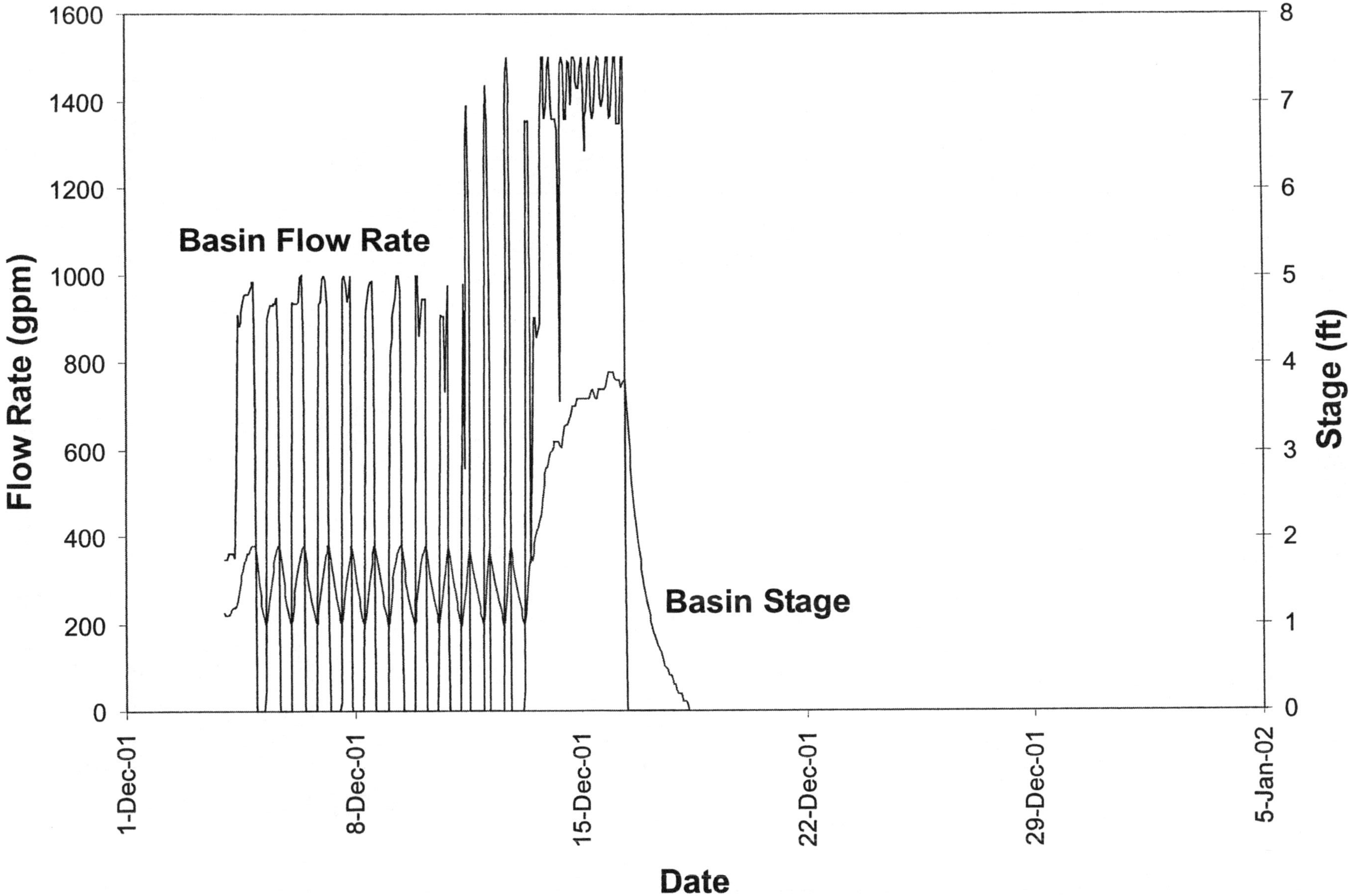

Figure F.1 (continued) Basin Flow Rate and Stage Versus Time, December 2001

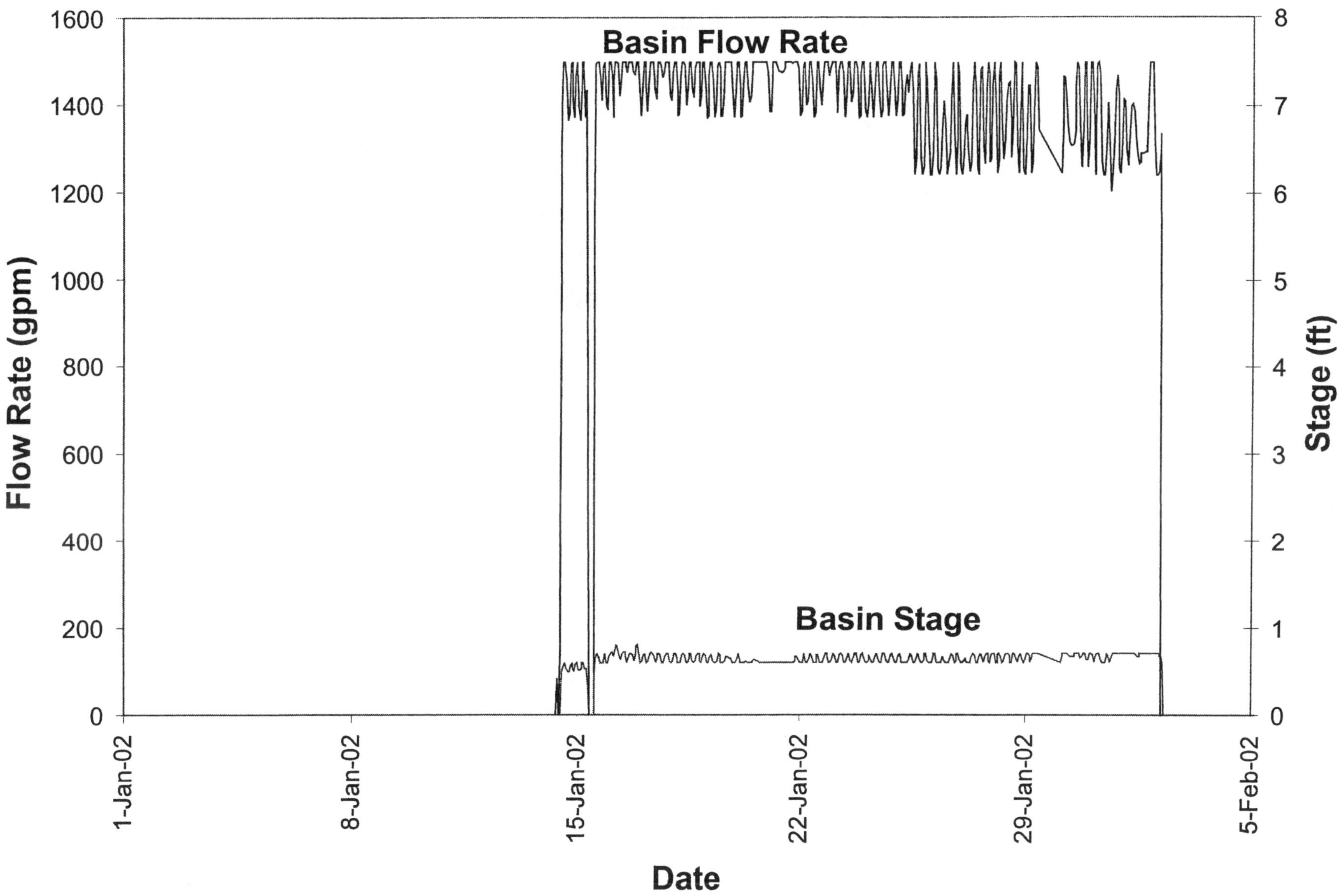

Figure F.1 (continued) Basin Flow Rate and Stage Versus Time, January 2002

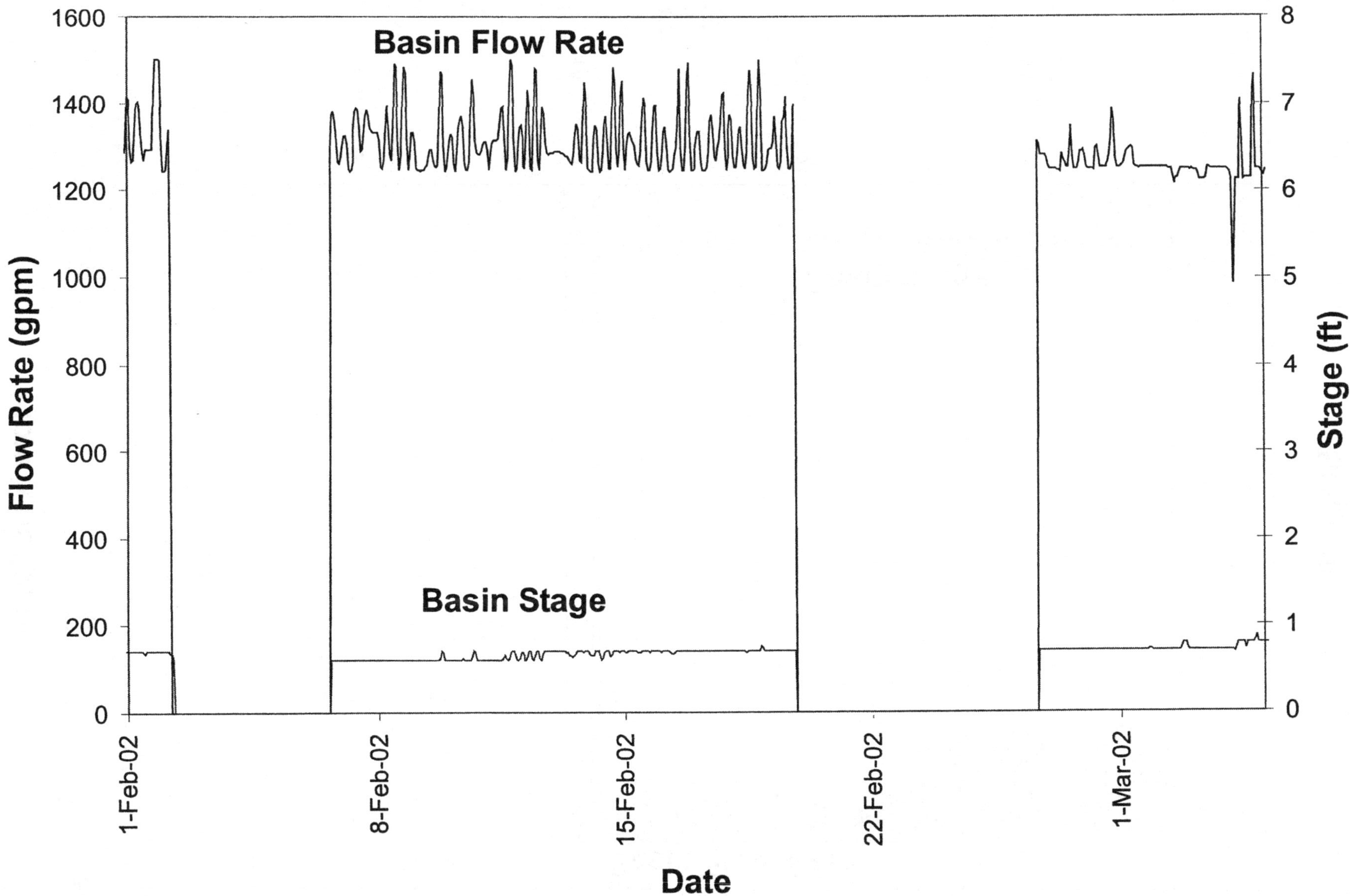

Figure F.1 (continued) Basin Flow Rate and Stage Versus Time, February 2002

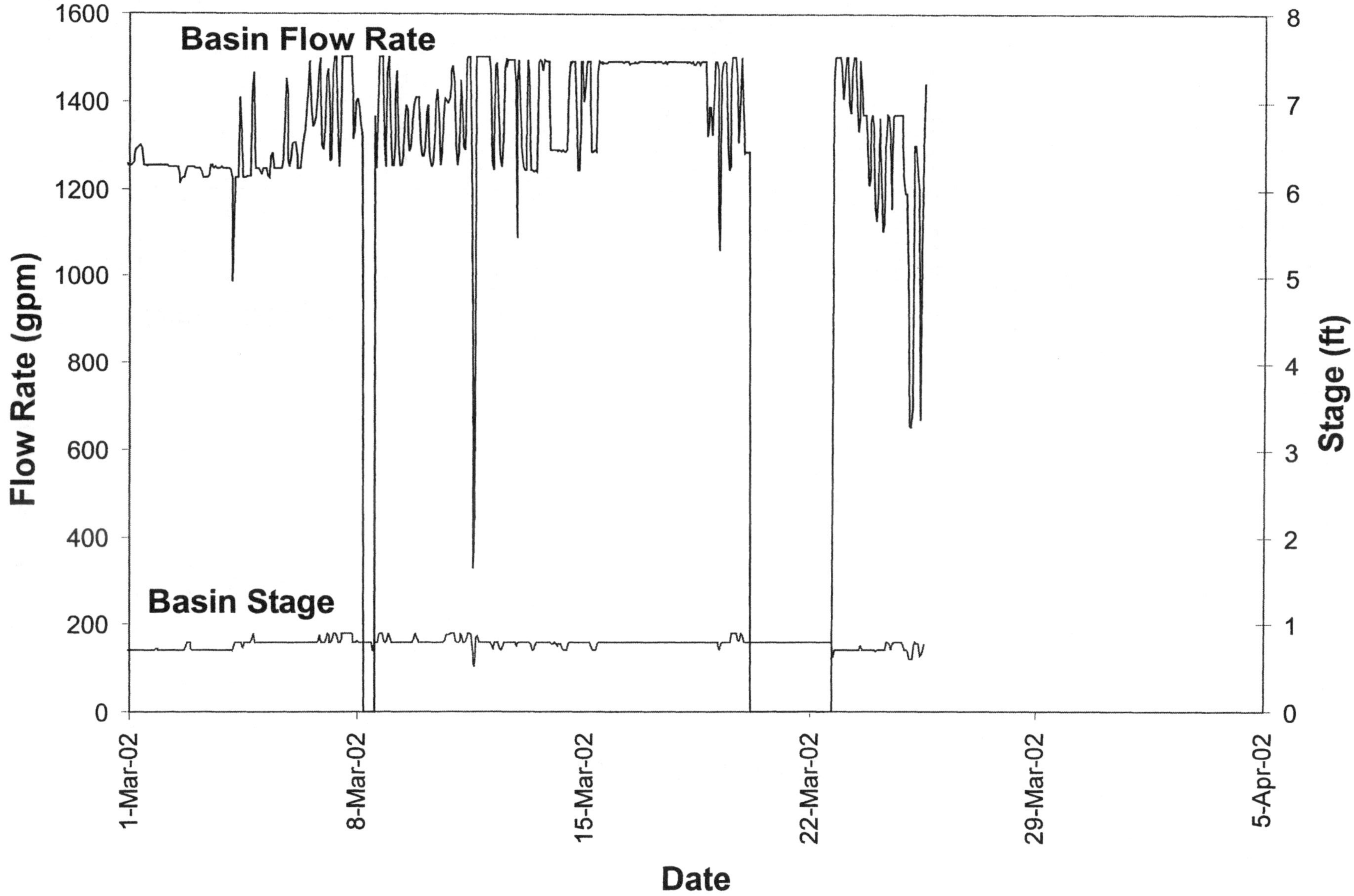

Figure F.1 (continued) Basin Flow Rate and Stage Versus Time, March 2002

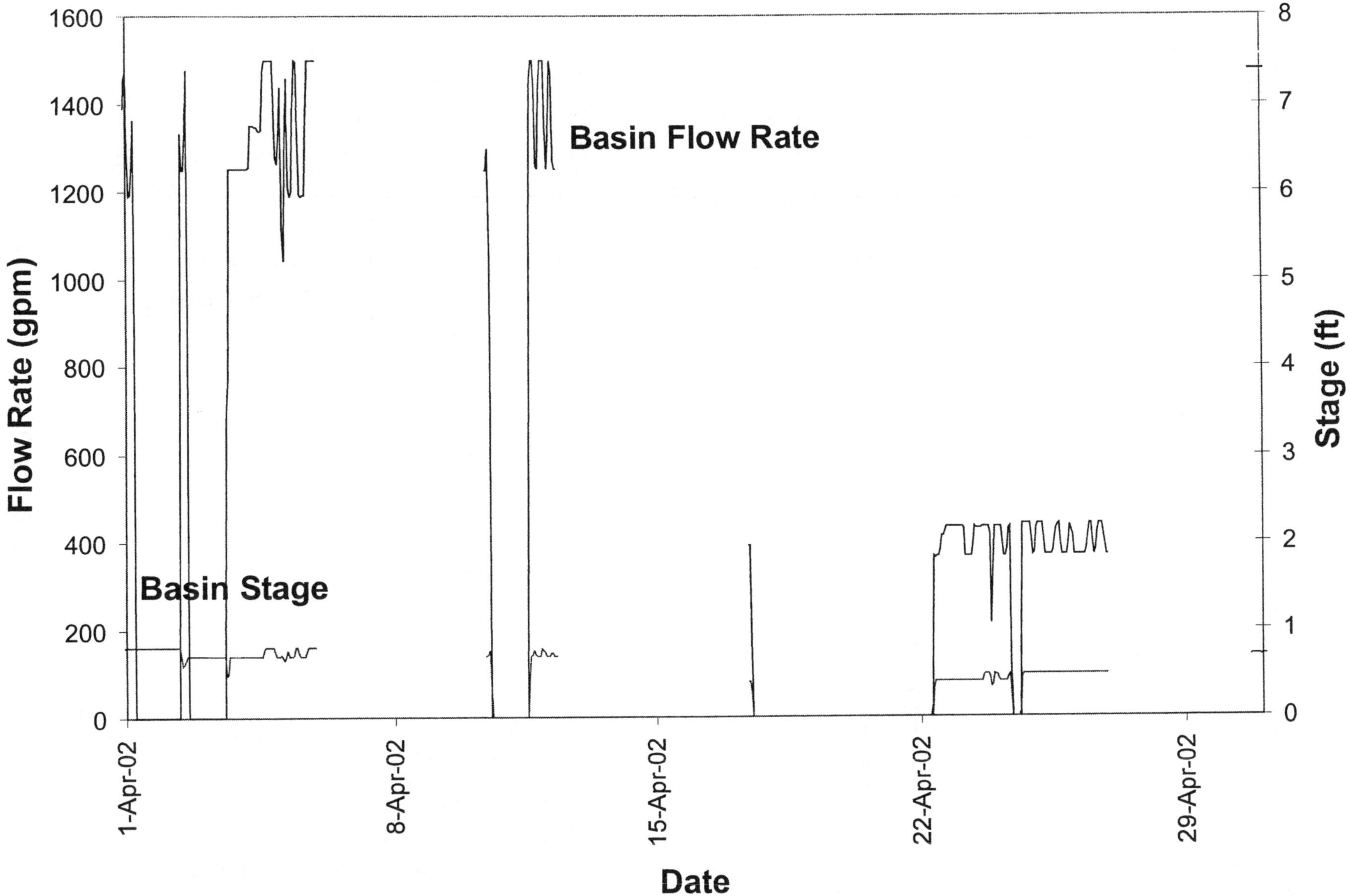

Figure F.1 (continued) Basin Flow Rate and Stage Versus Time, April 2002

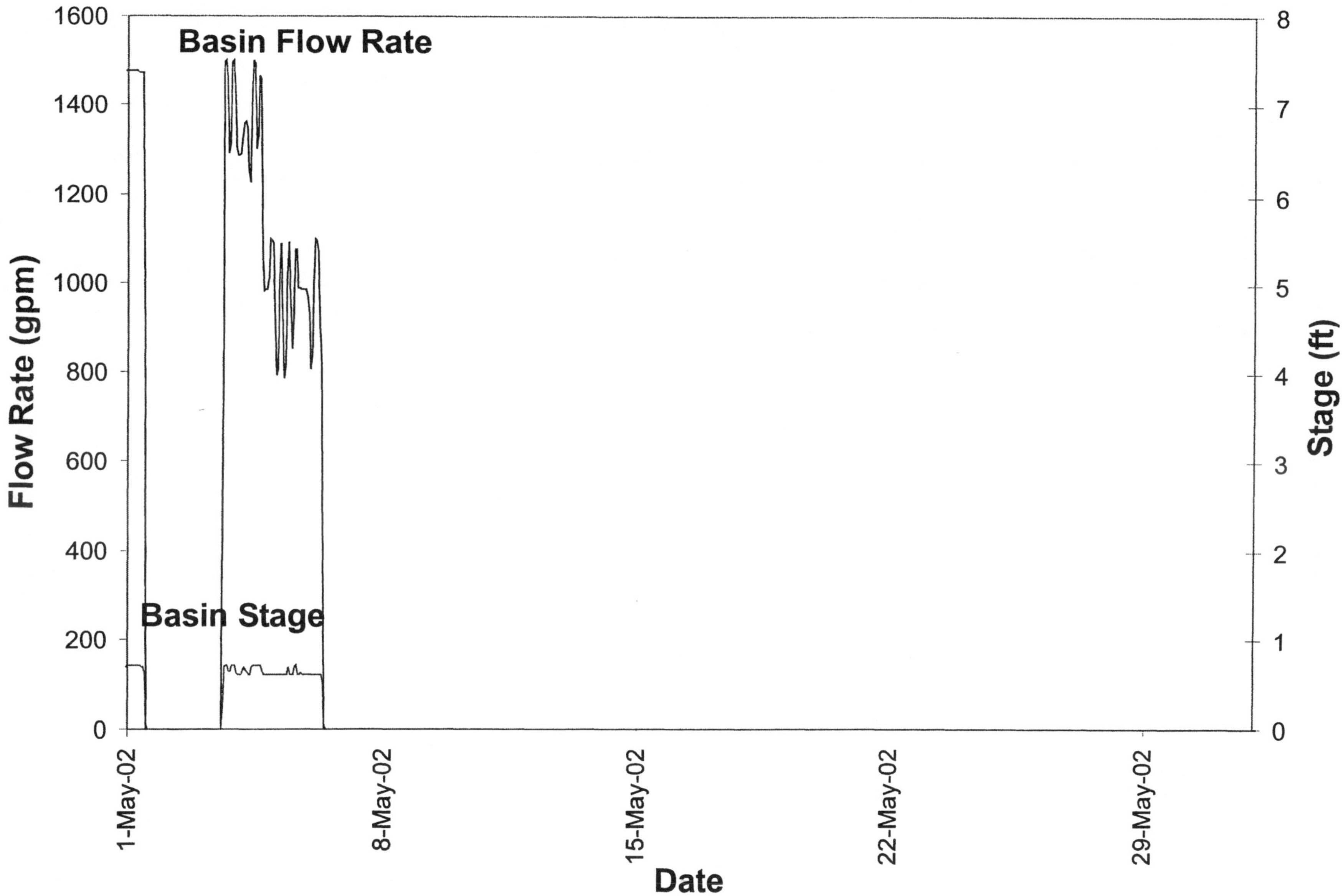

Figure F.1 (continued) Basin Flow Rate and Stage Versus Time, May 2002

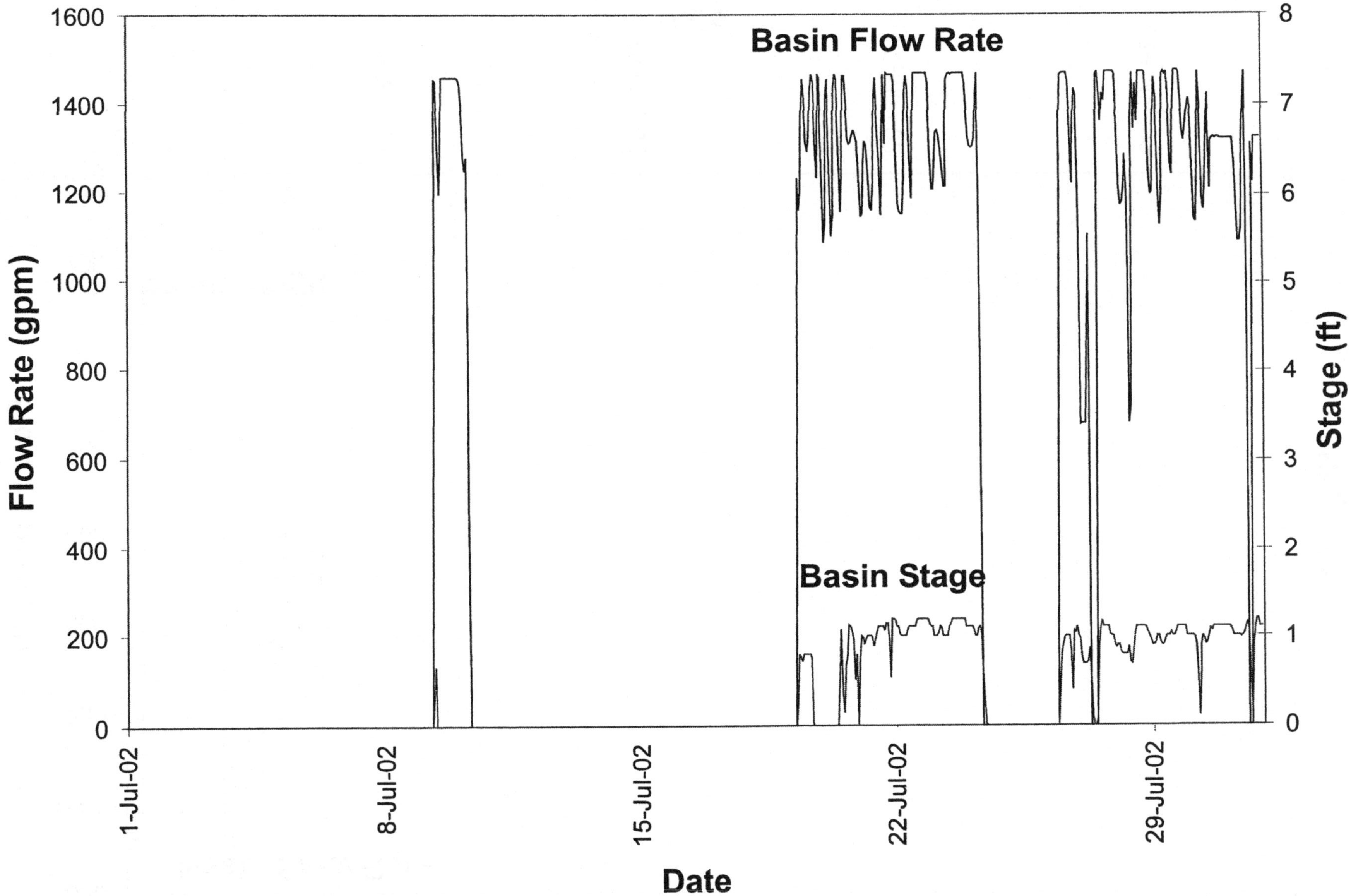

Figure F.1 (continued) Basin Flow Rate and Stage Versus Time, July 2002

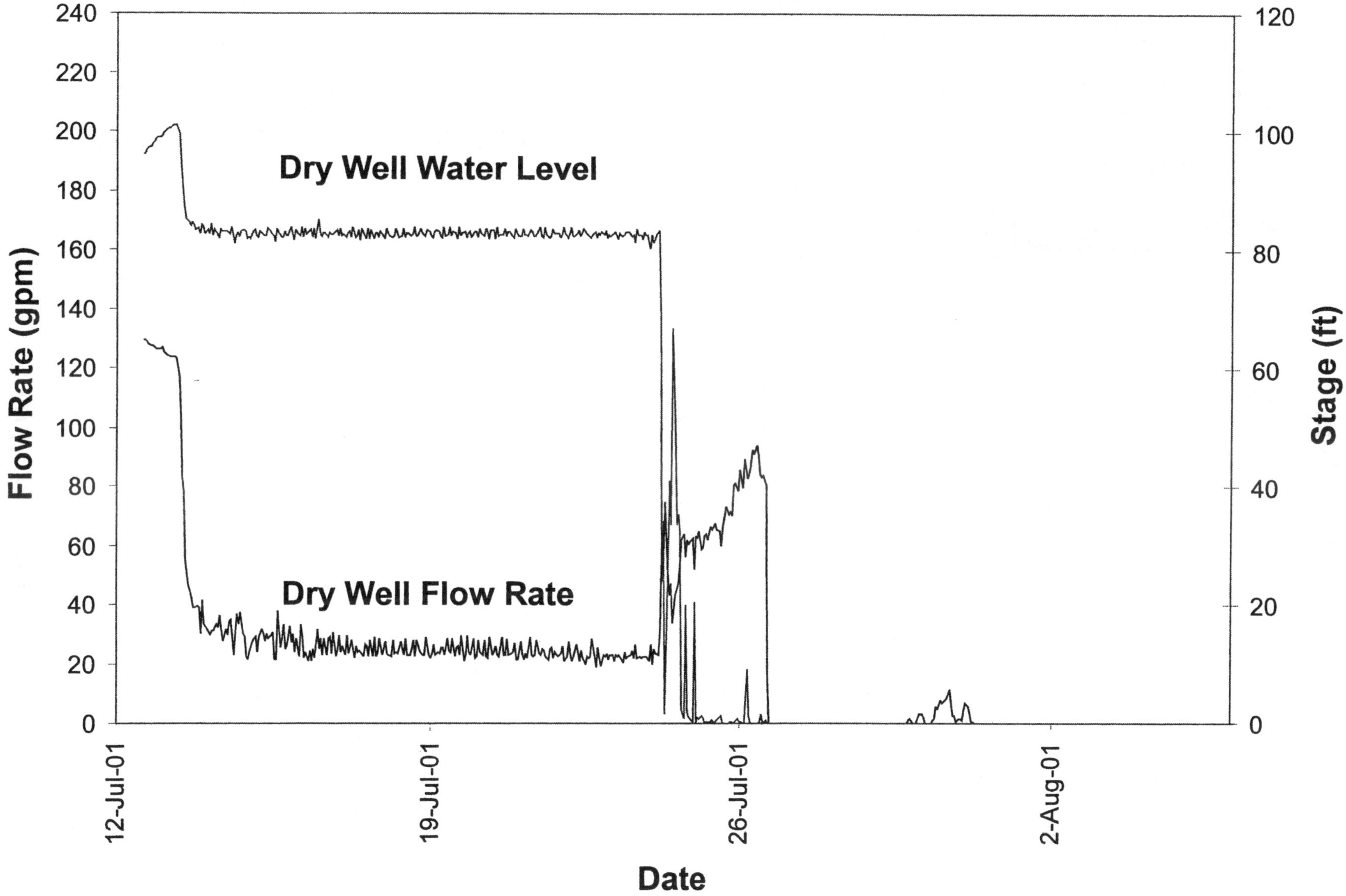

Figure F.2 Dry well flow rate and stage versus time, July 2001

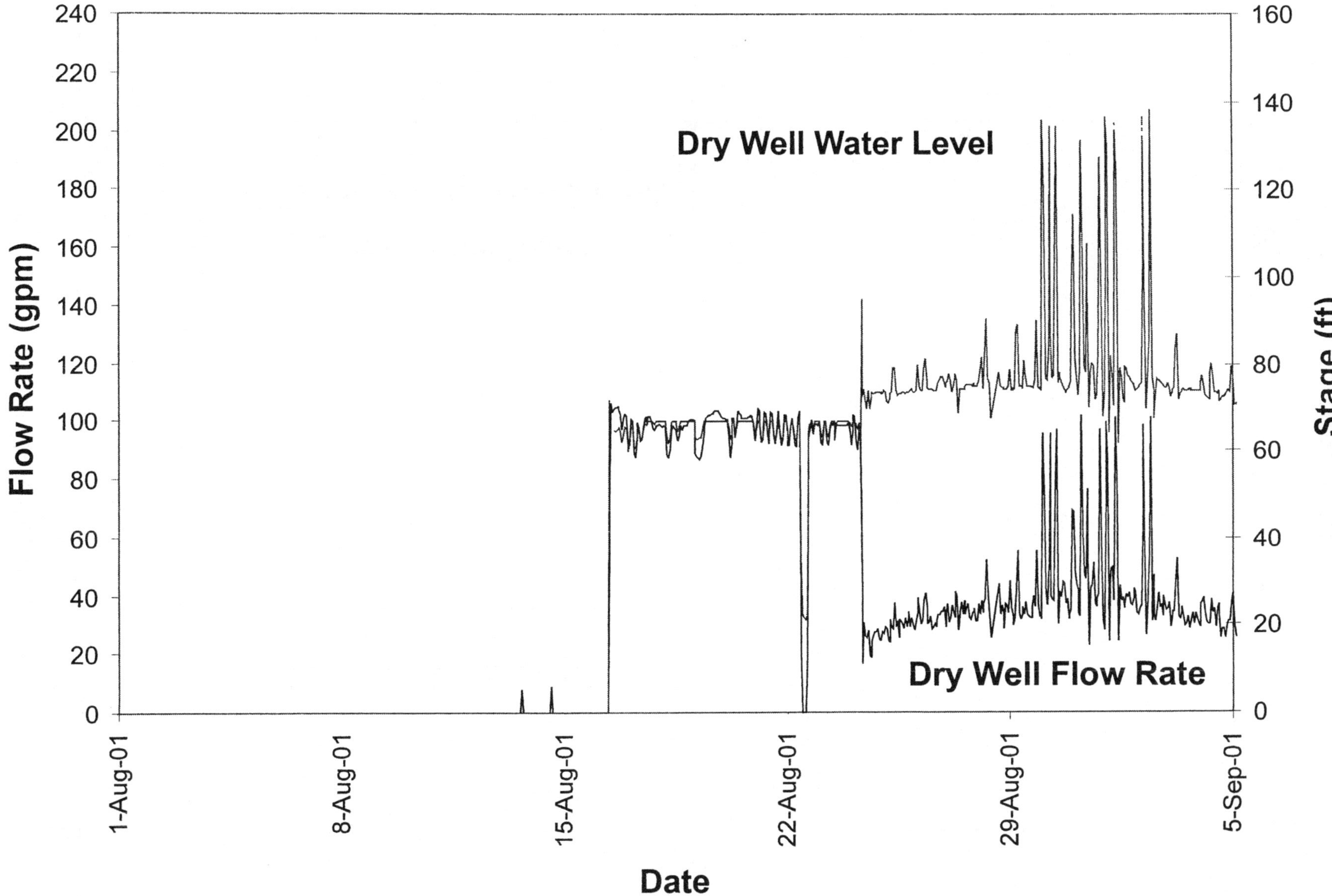

Figure F.2 (continued) Dry well flow rate and stage versus time, August 2001

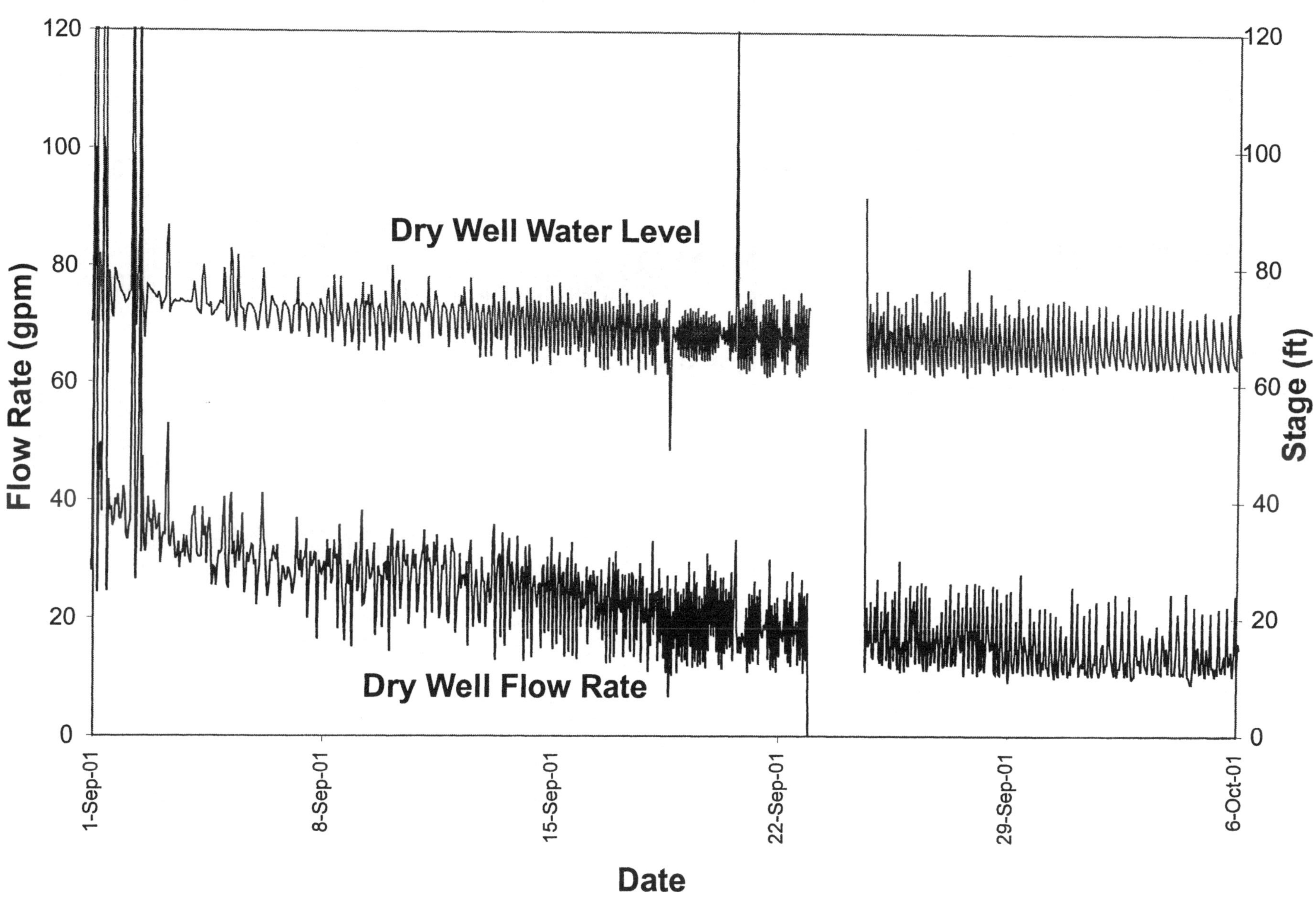

Figure F.2 (continued) Dry well flow rate and stage versus time, September 2001

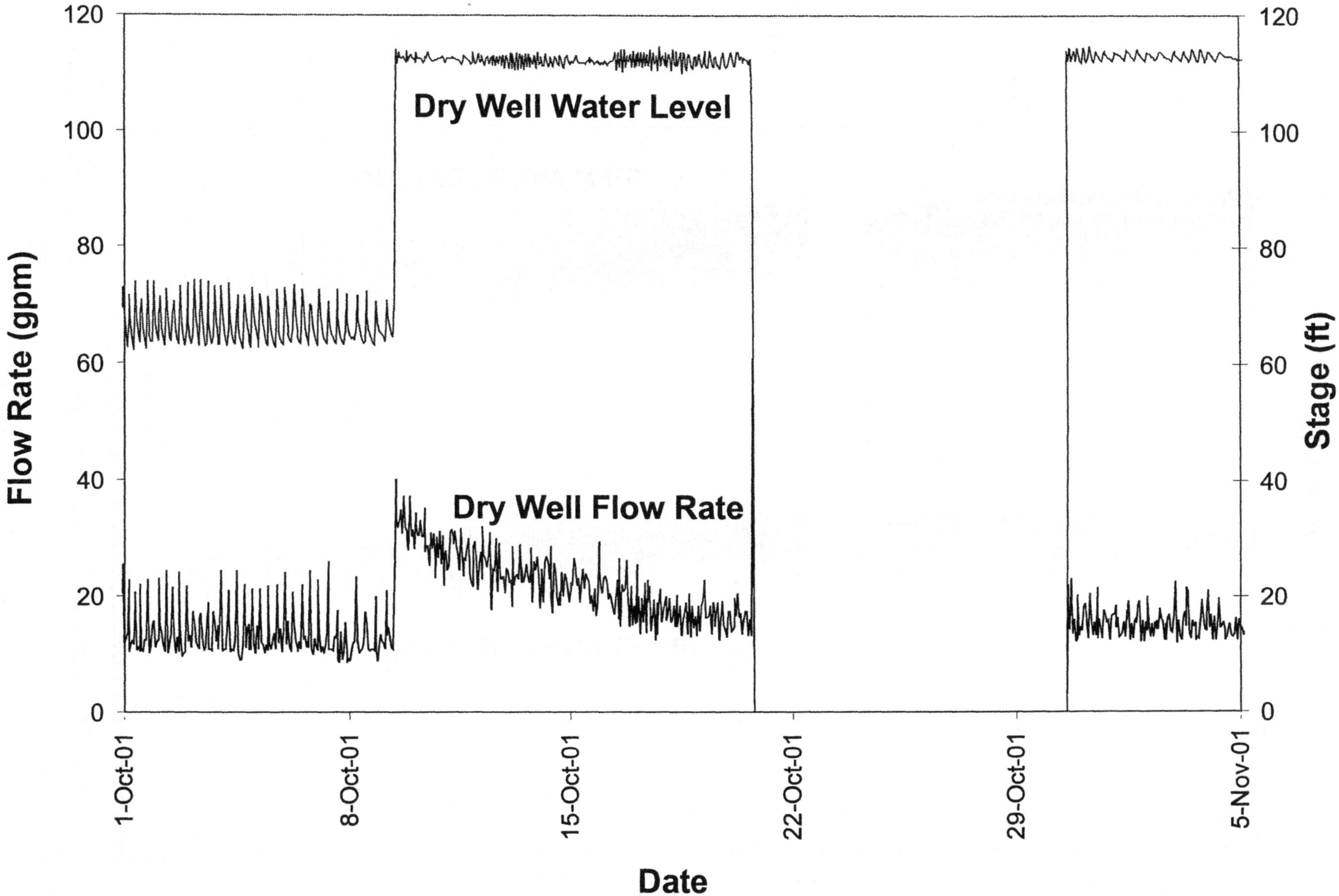

Figure F.2 (continued) Dry well flow rate and stage versus time, October 2001

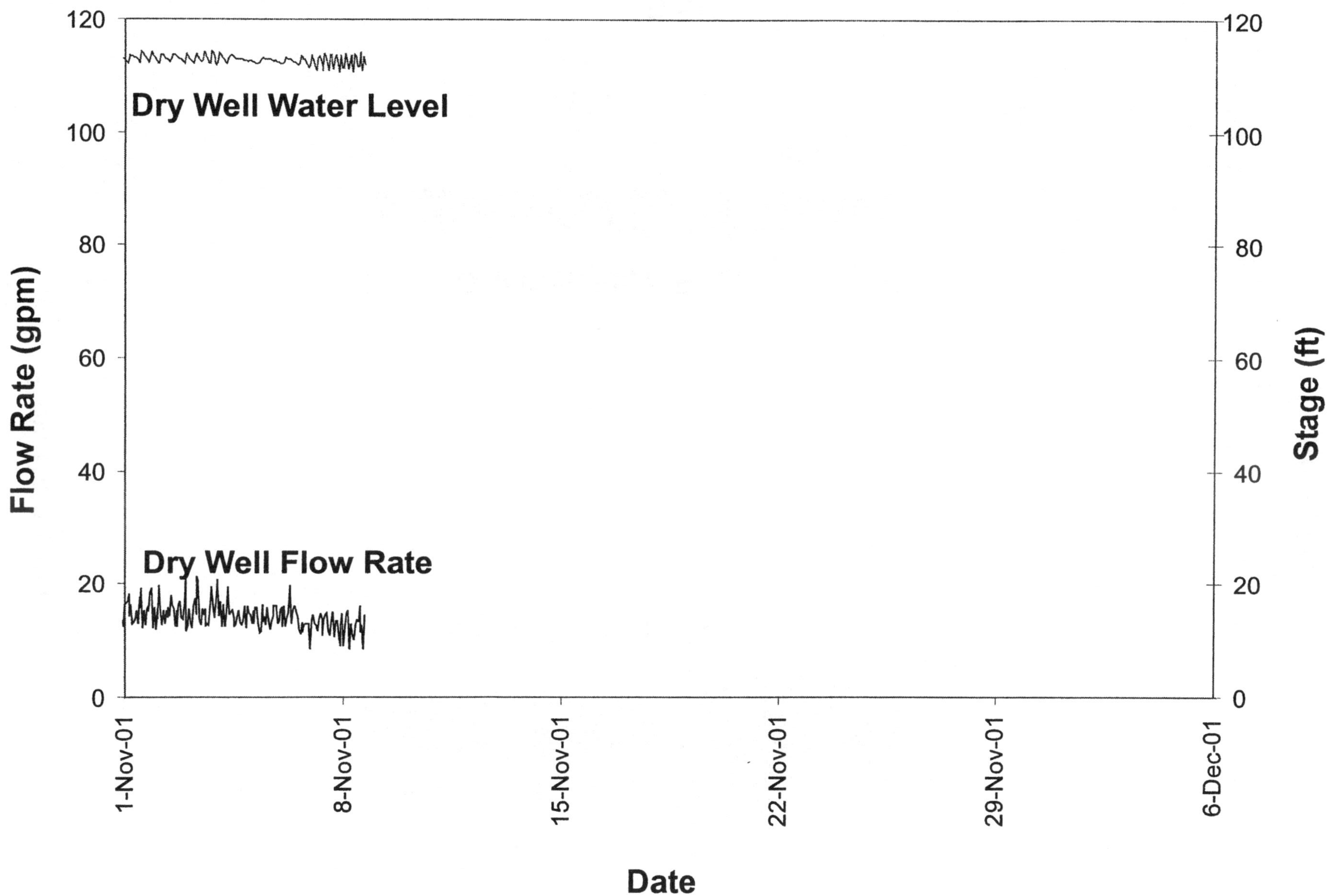

Figure F.2 (continued) Dry well flow rate and stage versus time, November 2001

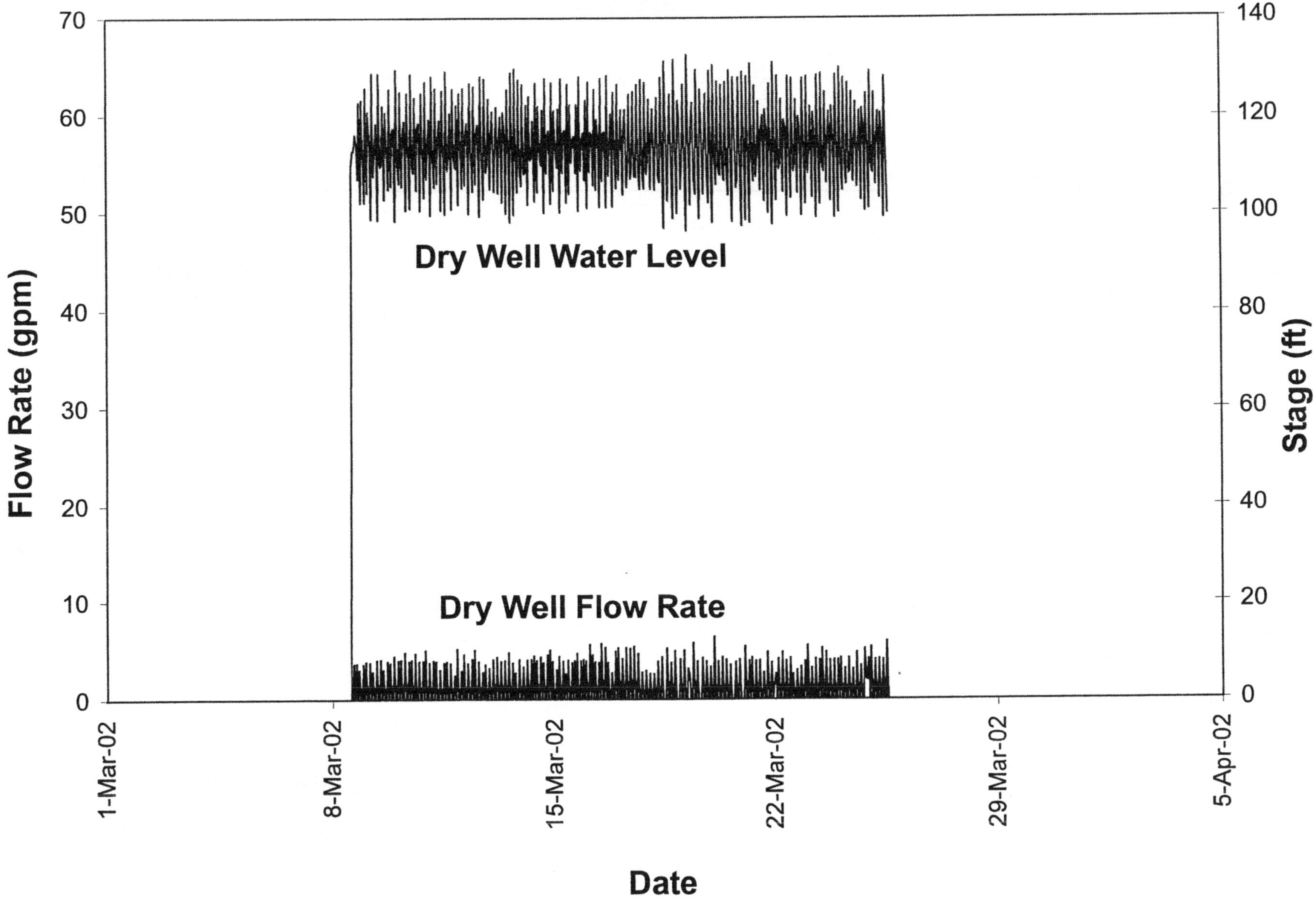

Figure F.2 (continued) Dry well flow rate and stage versus time, March 2002

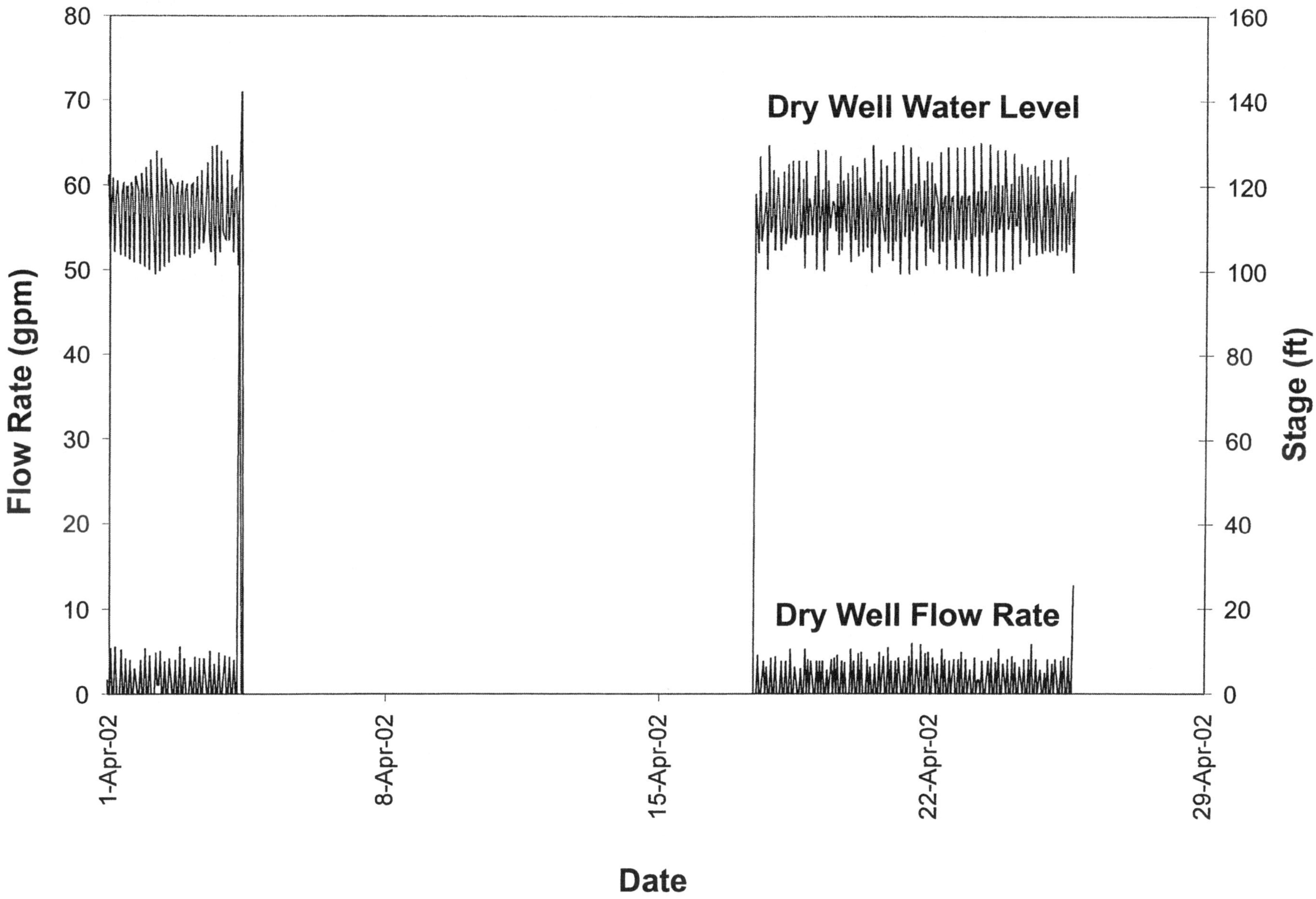

Figure F.2 (continued) Dry well flow rate and stage versus time, April 2002

APPENDIX G
THERMOCOUPLE DATA

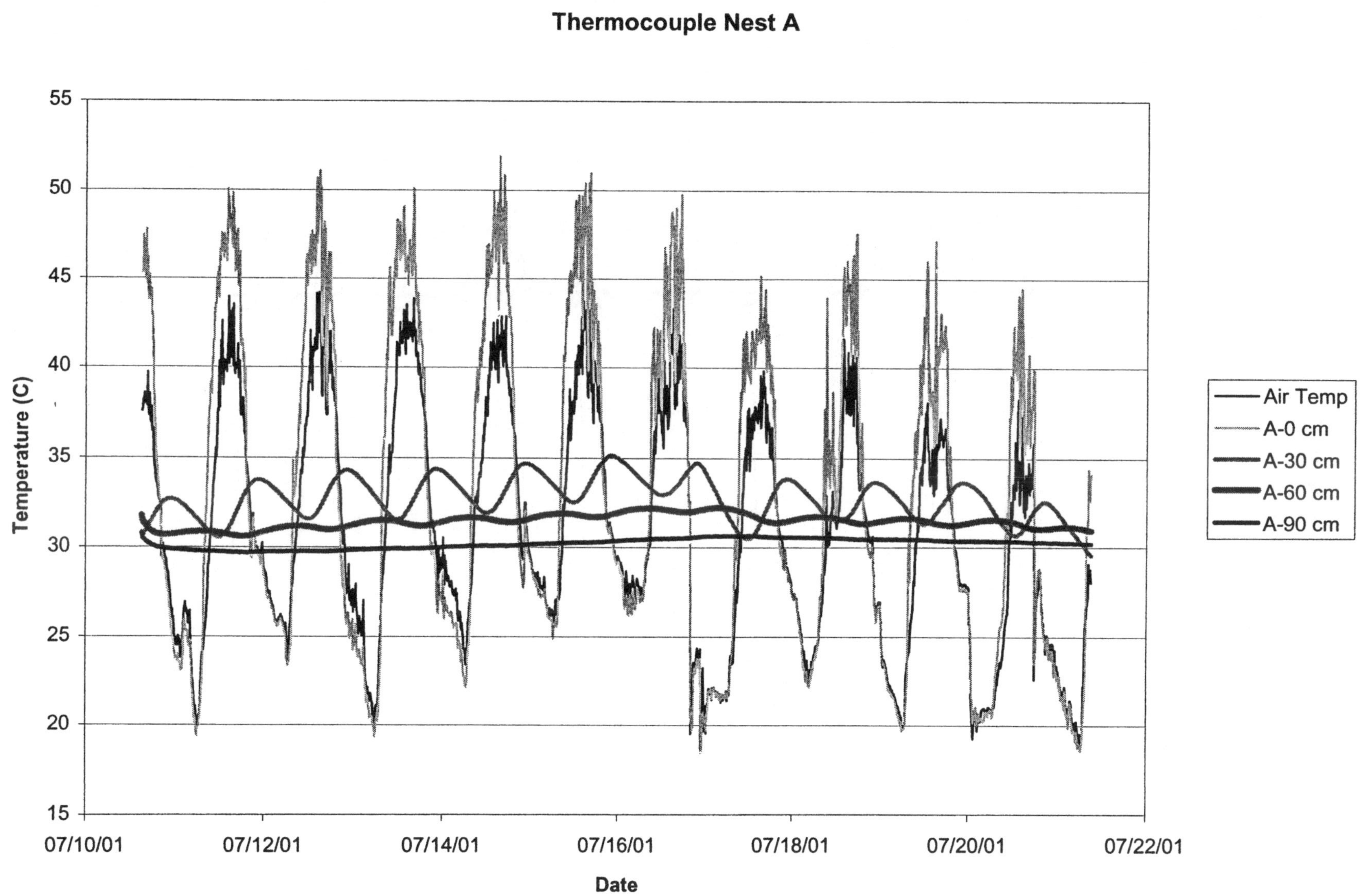

Figure G.1 Thermocouple data 7/10/01 through 8/21/01

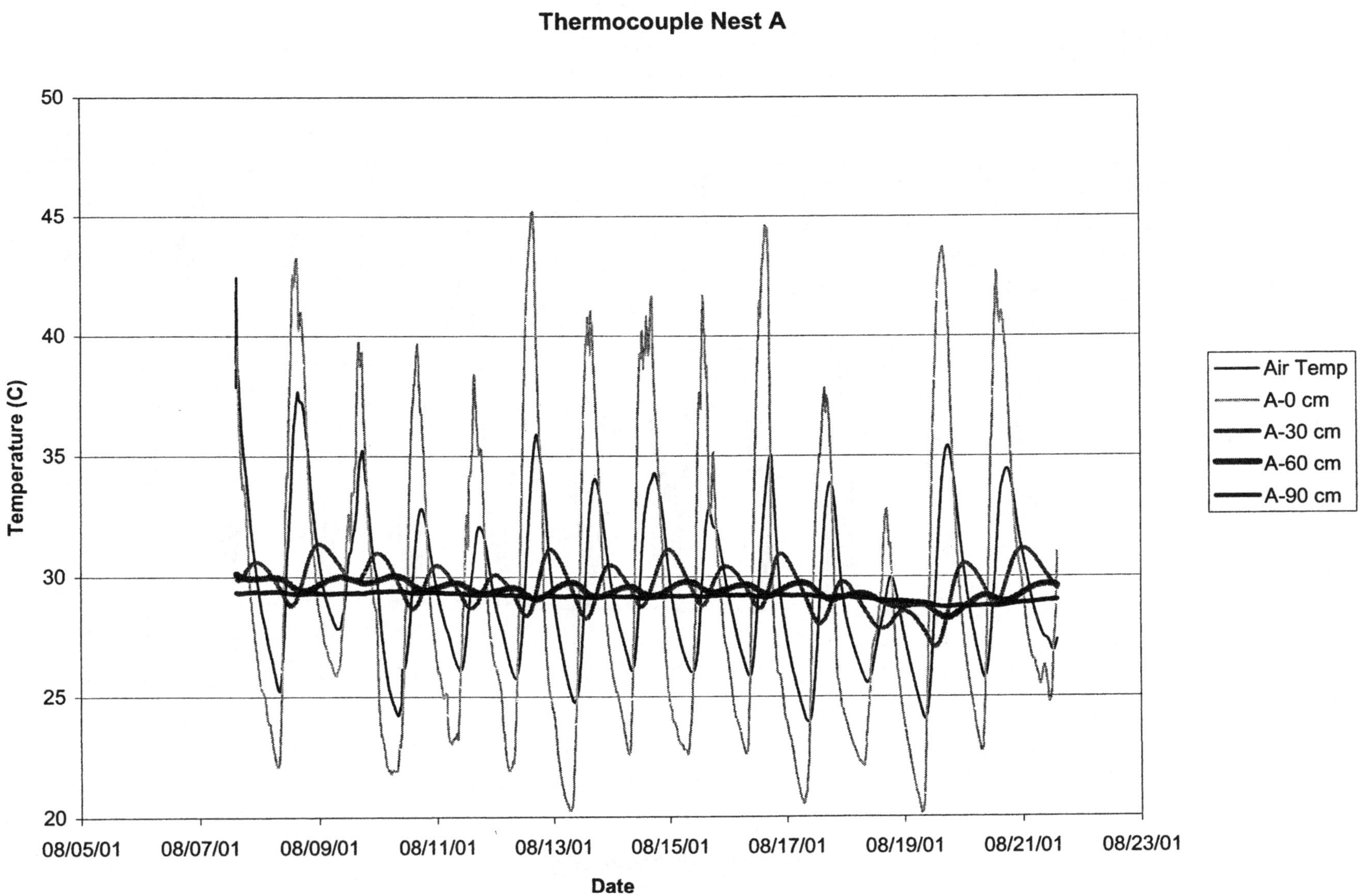

Figure G.1 (continued) Thermocouple data 7/10/01 through 8/21/01

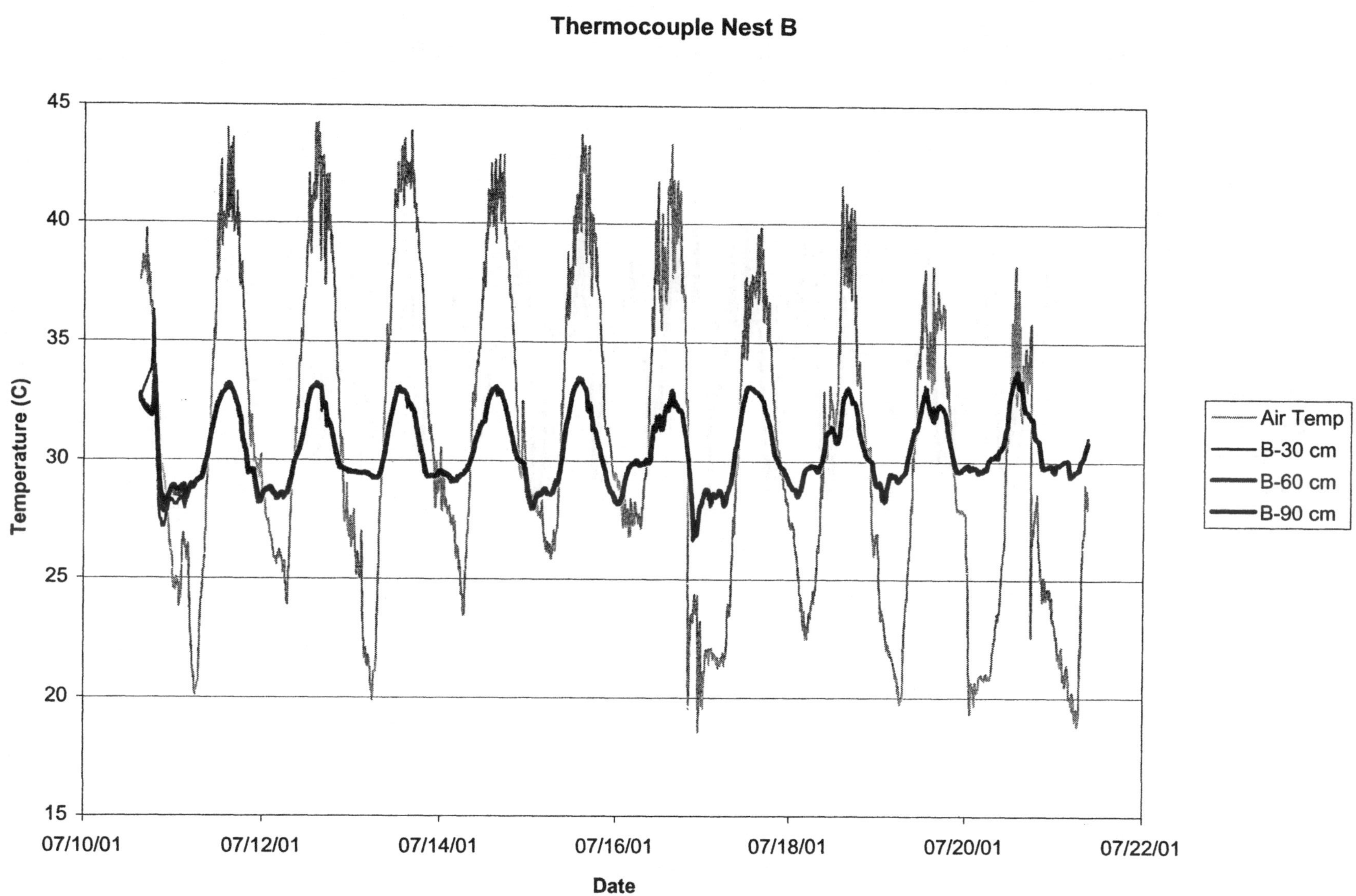

Figure G.1 (continued) Thermocouple data 7/10/01 through 8/21/01

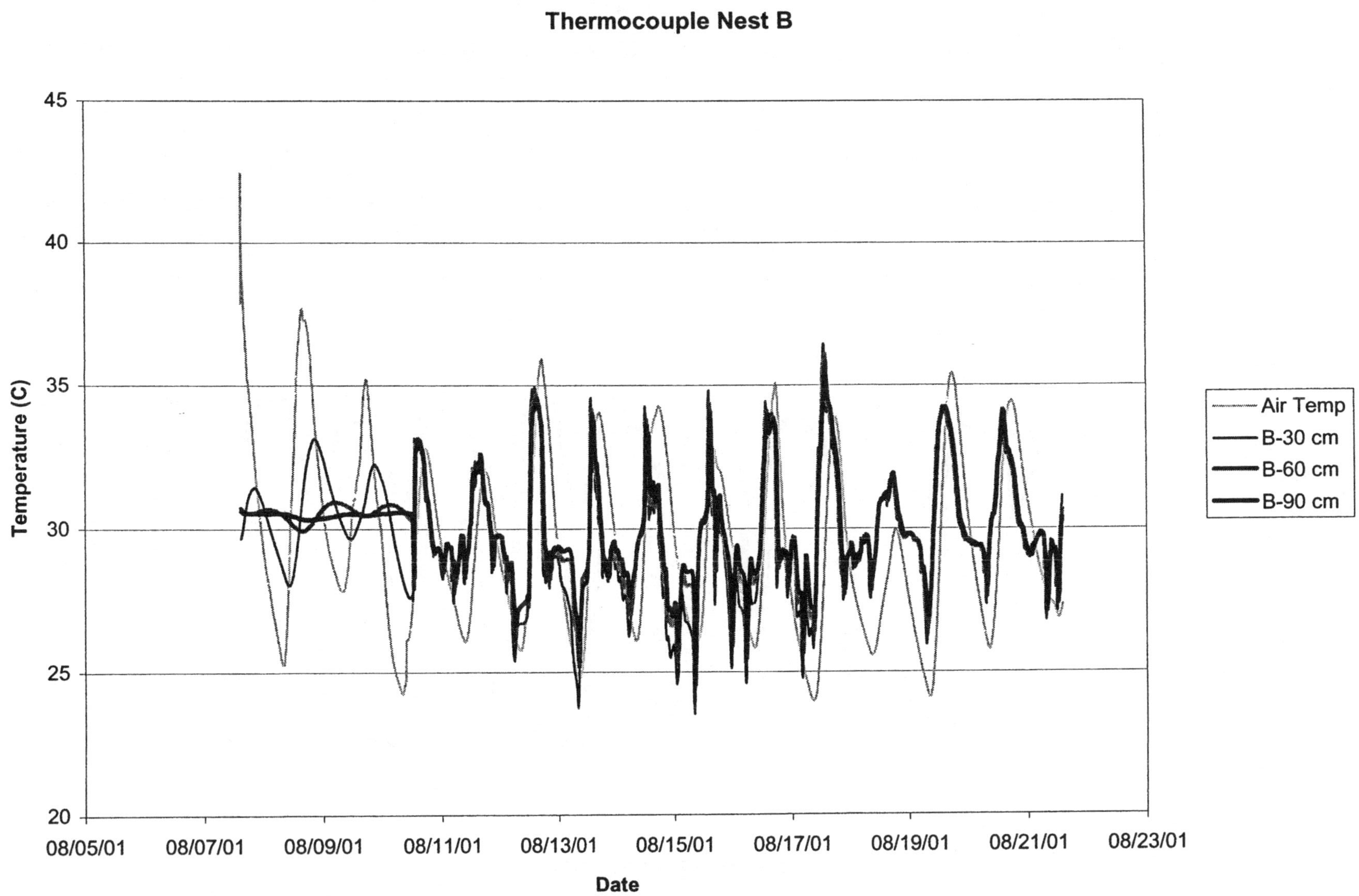

Figure G.1 (continued) Thermocouple data 7/10/01 through 8/21/01

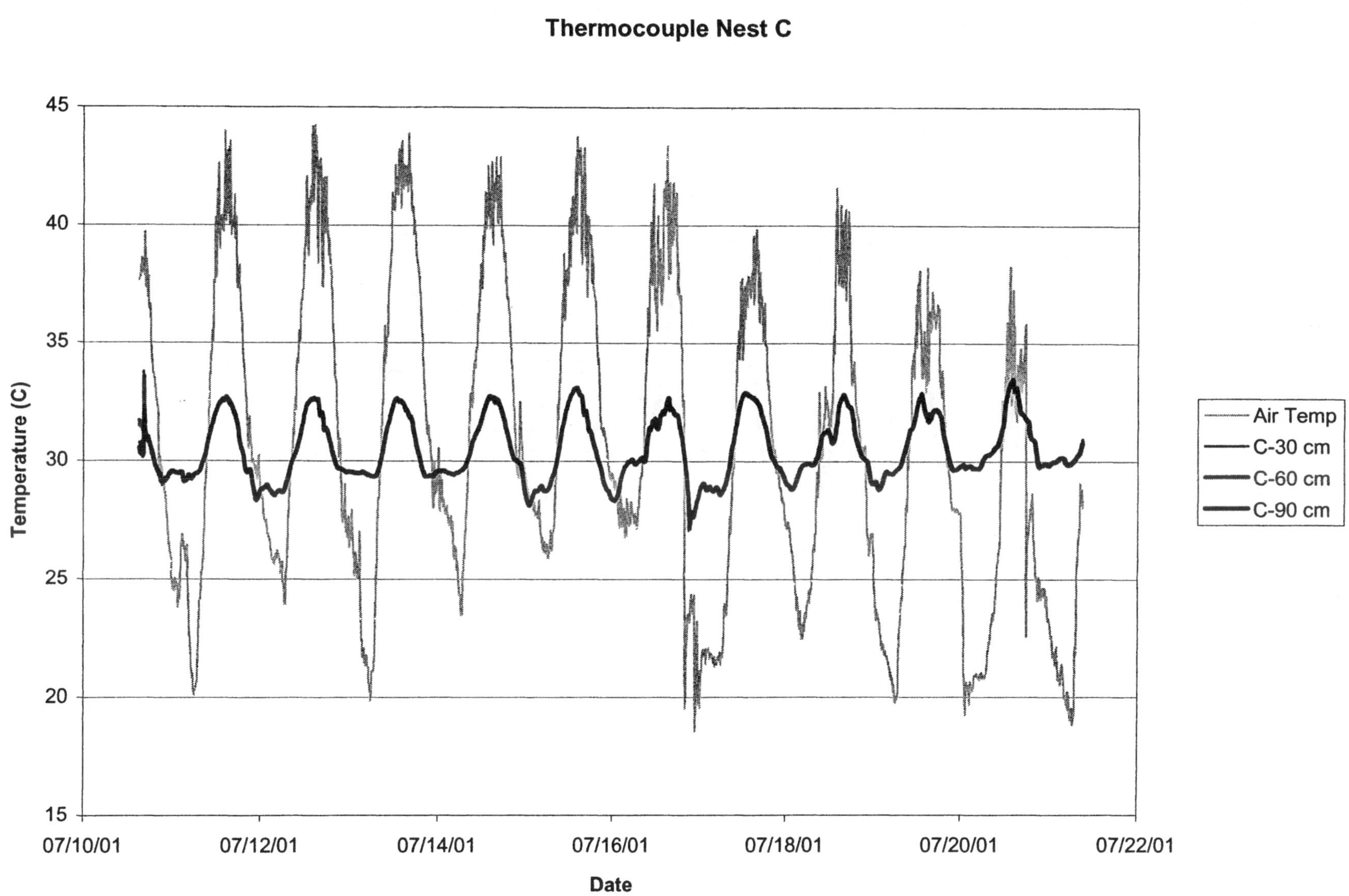

Figure G.1 (continued) Thermocouple data 7/10/01 through 8/21/01

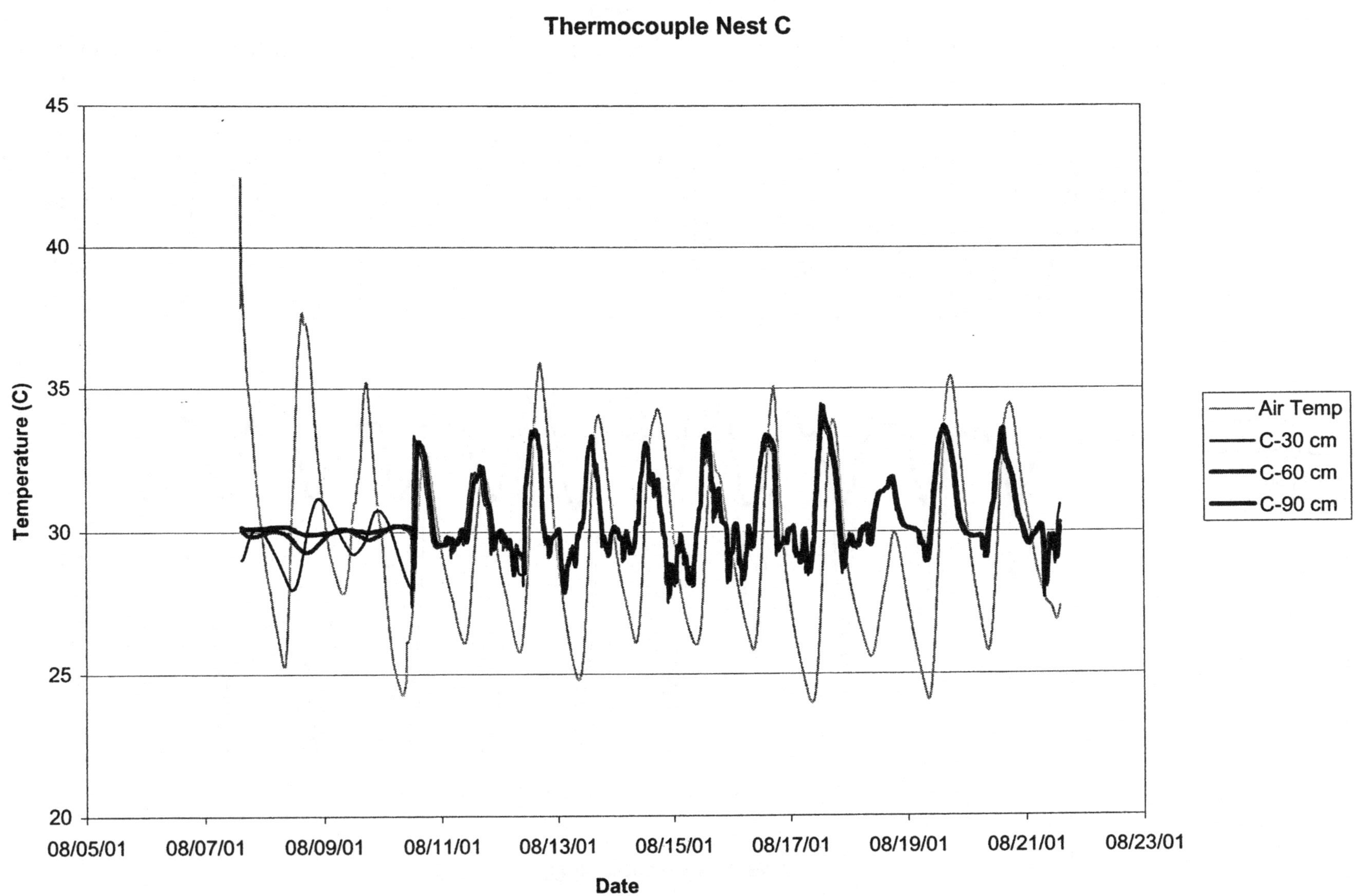

Figure G.1 (continued) Thermocouple data 7/10/01 through 8/21/01

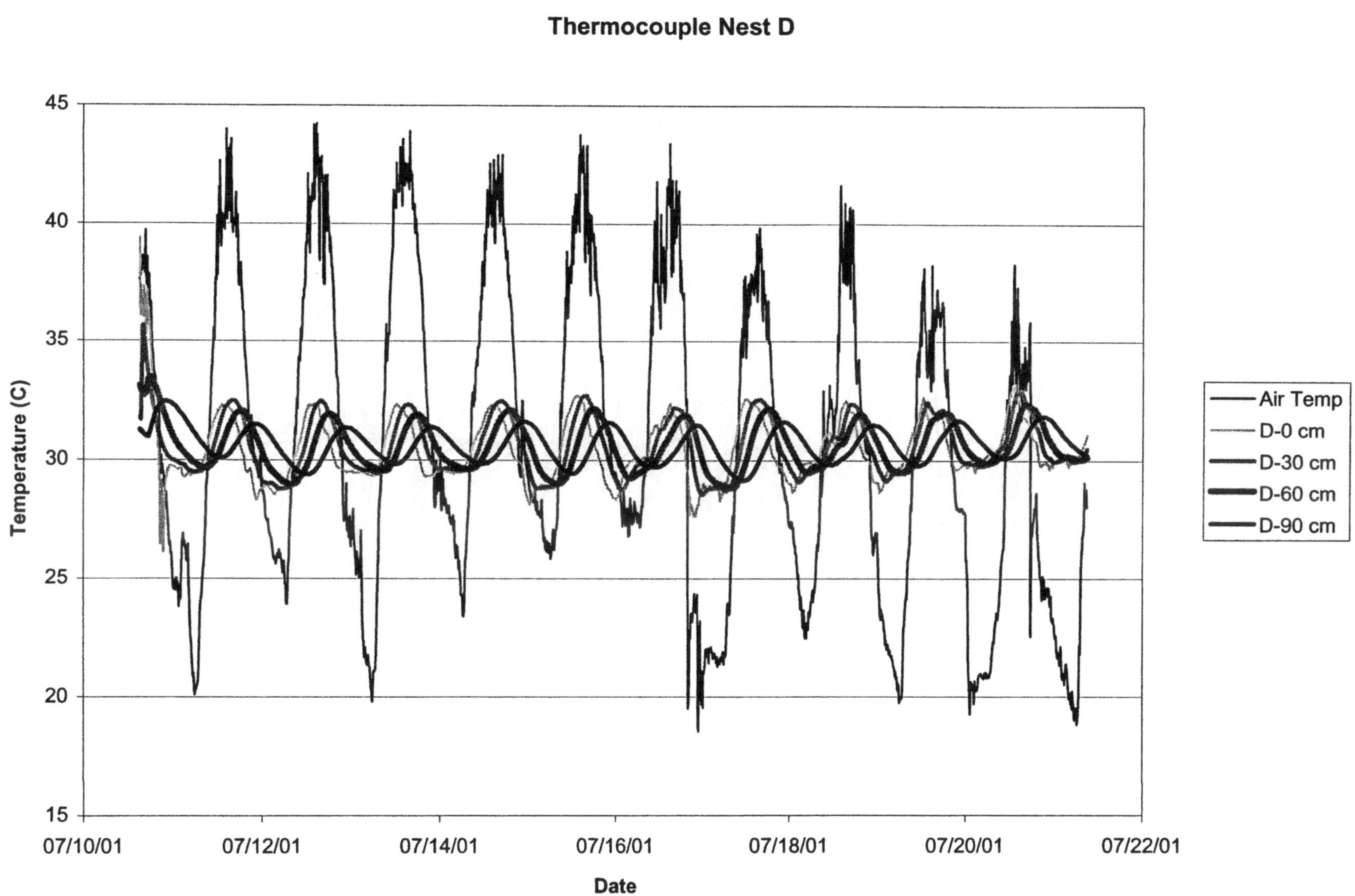

Figure G.1 (continued) Thermocouple data 7/10/01 through 8/21/01

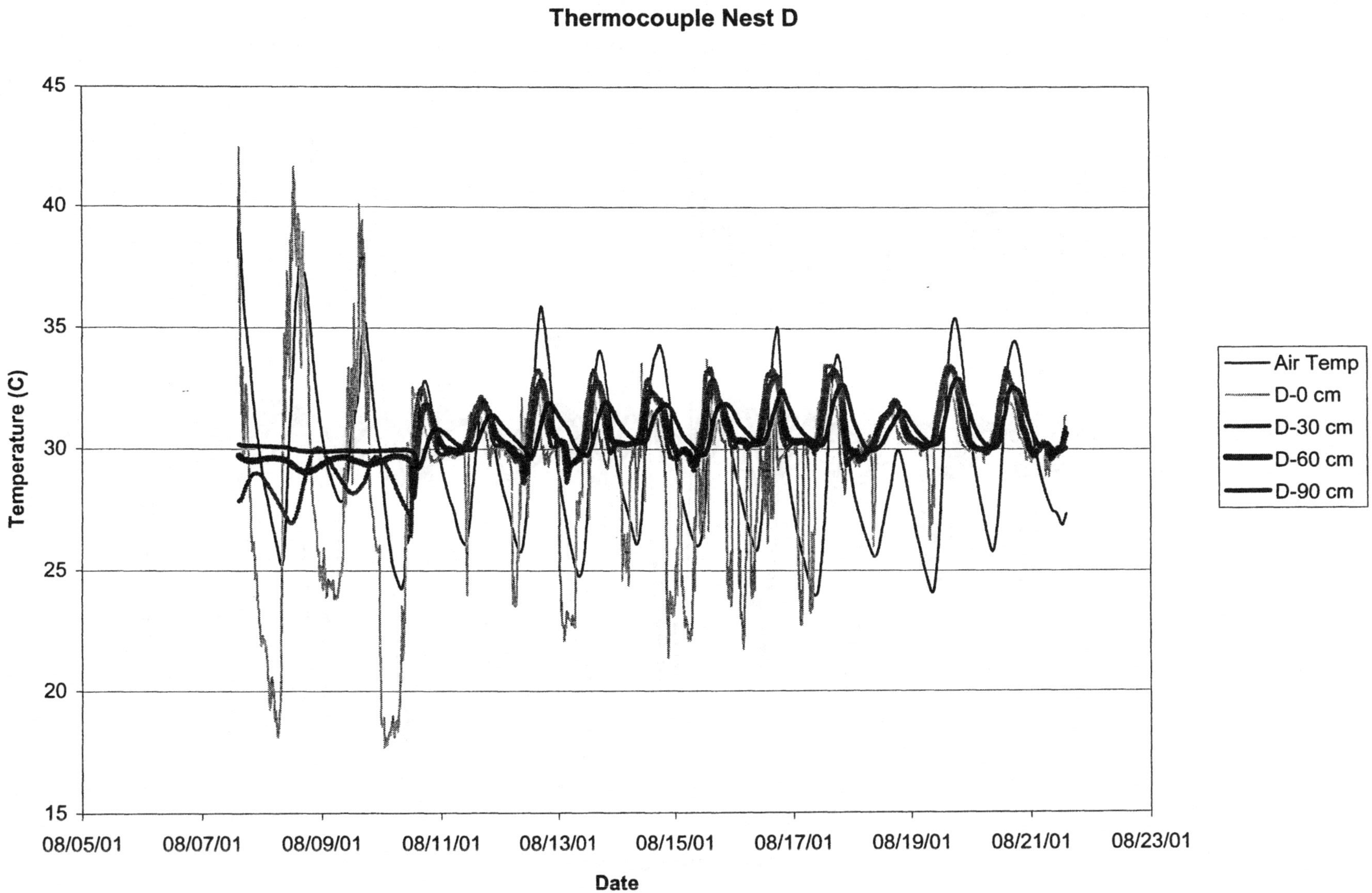

Figure G.1 (continued) Thermocouple data 7/10/01 through 8/21/01

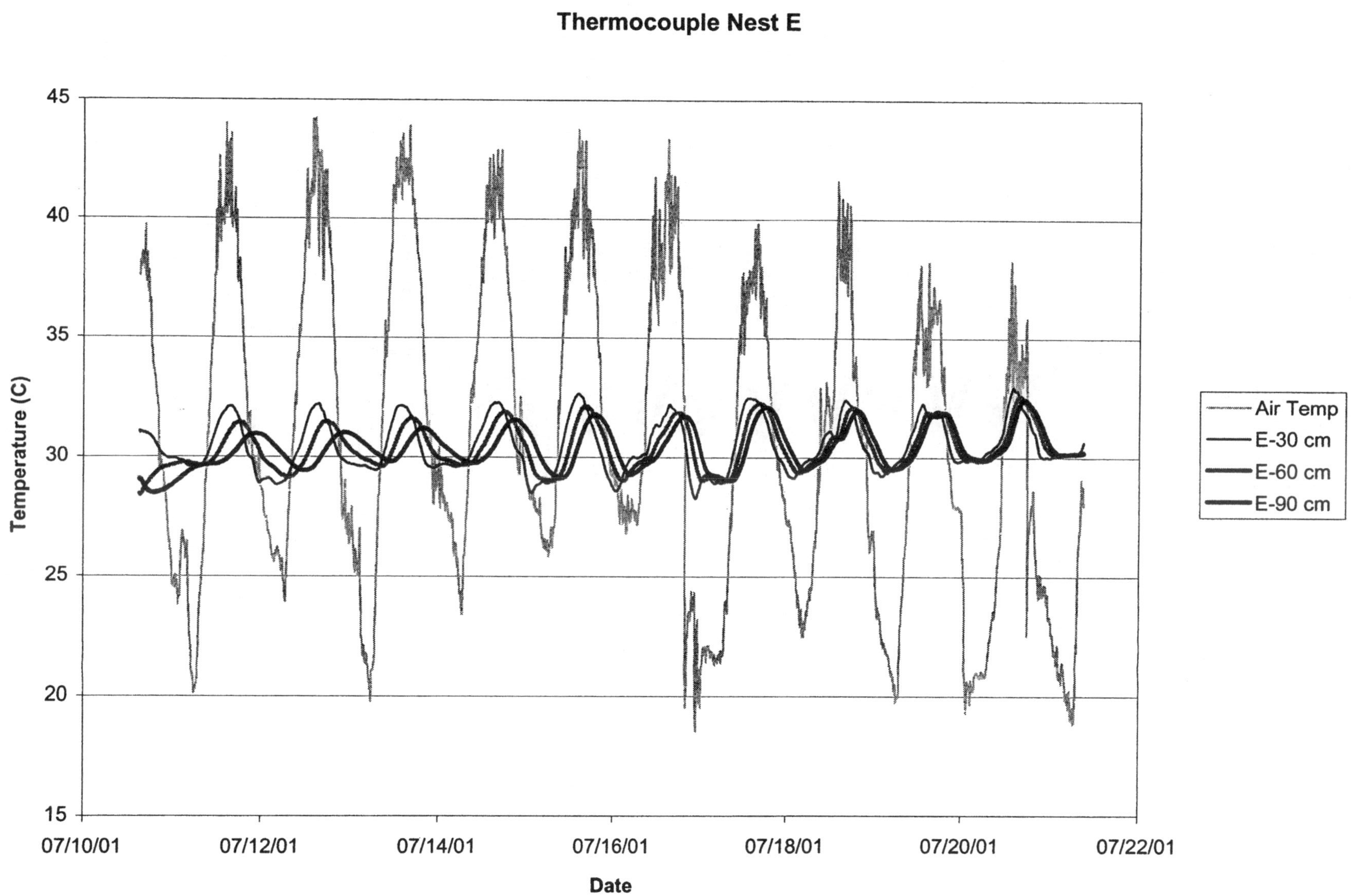

Figure G.1 (continued) Thermocouple data 7/10/01 through 8/21/01

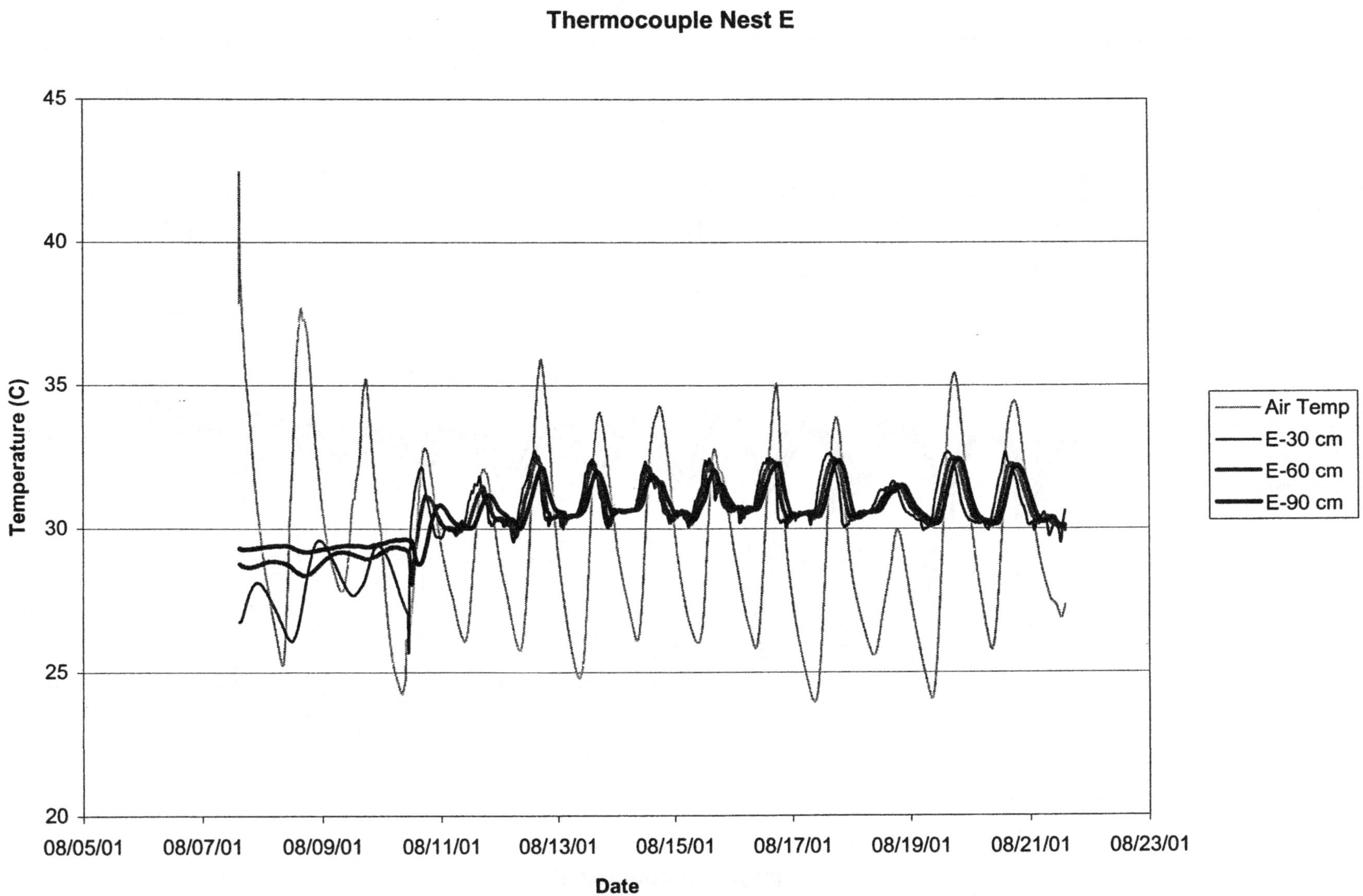

Figure G.1 (continued) Thermocouple data 7/10/01 through 8/21/01

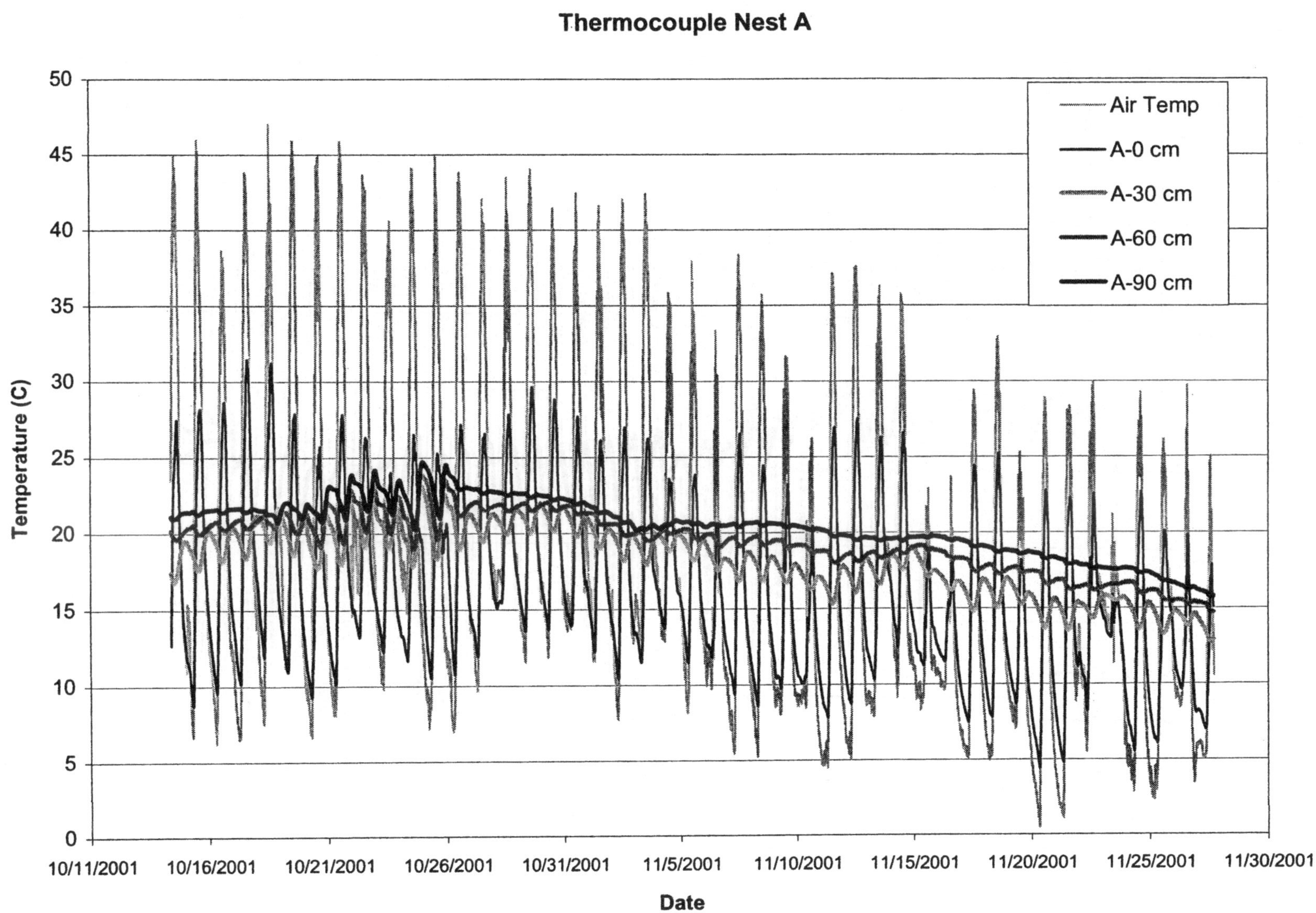

Figure G.2 Thermocouple data 10/14/01 through 11/27/02

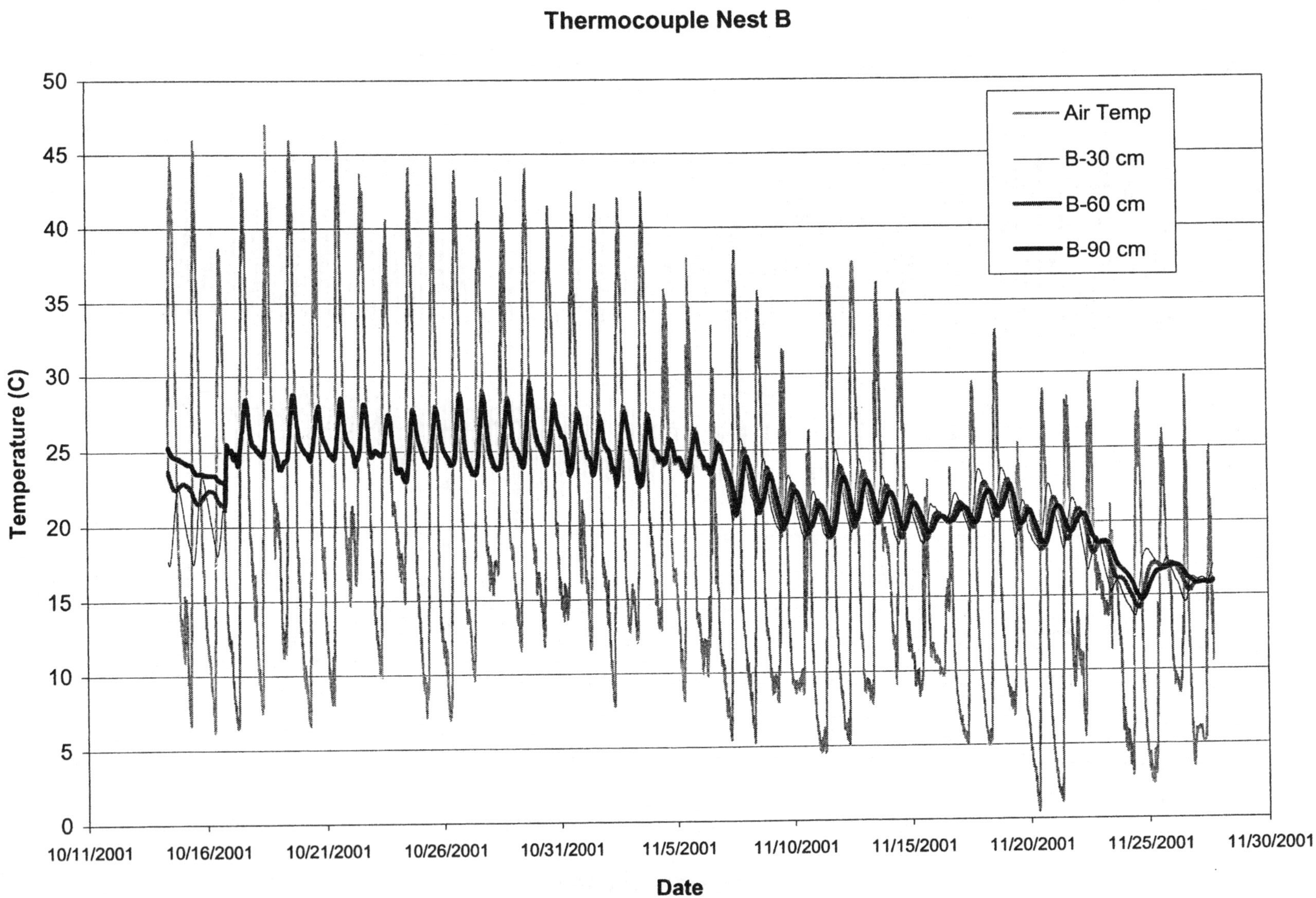

Figure G.2 (continued) Thermocouple data 10/14/01 through 11/27/02

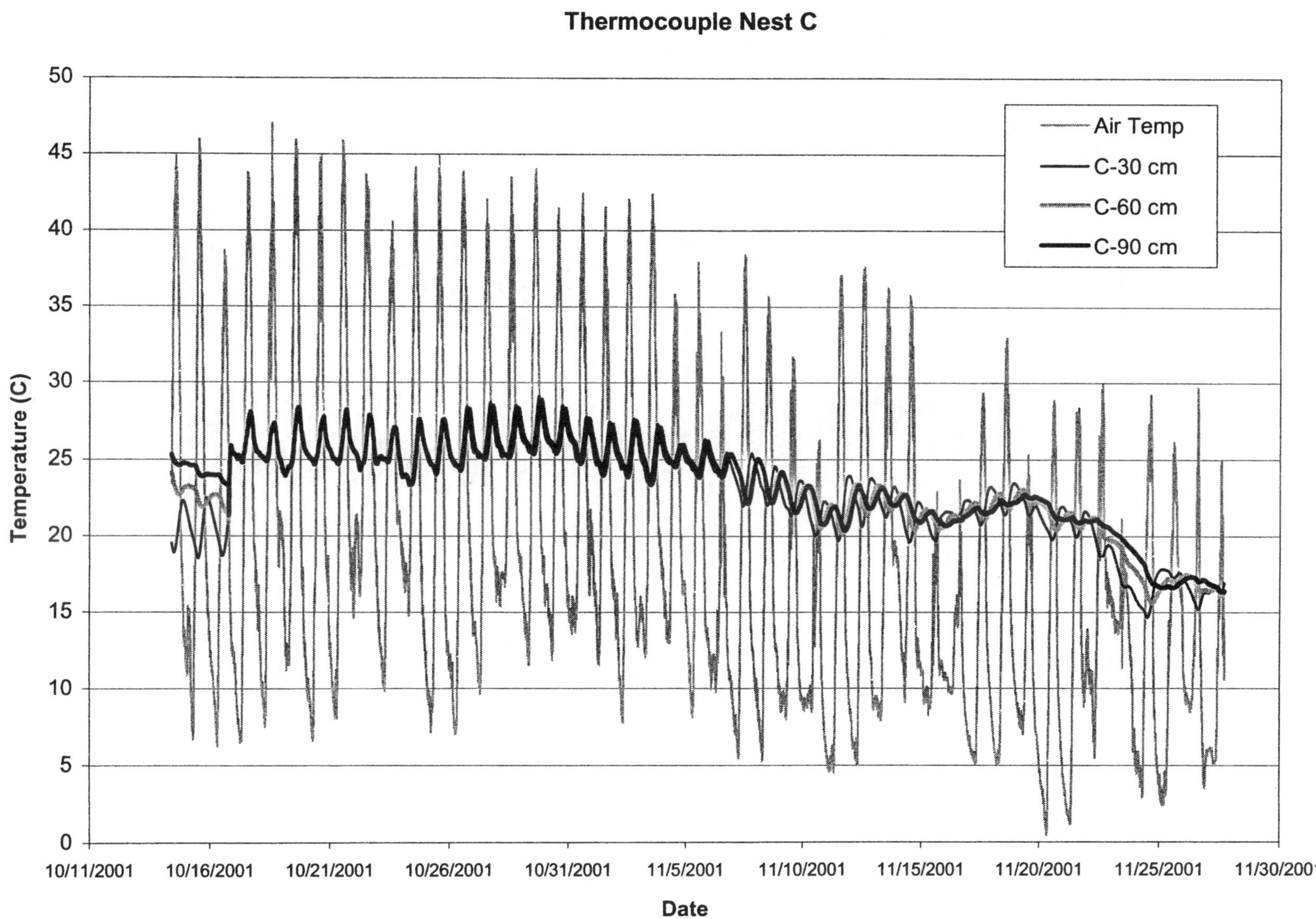

Figure G.2 (continued) Thermocouple data 10/14/01 through 11/27/02

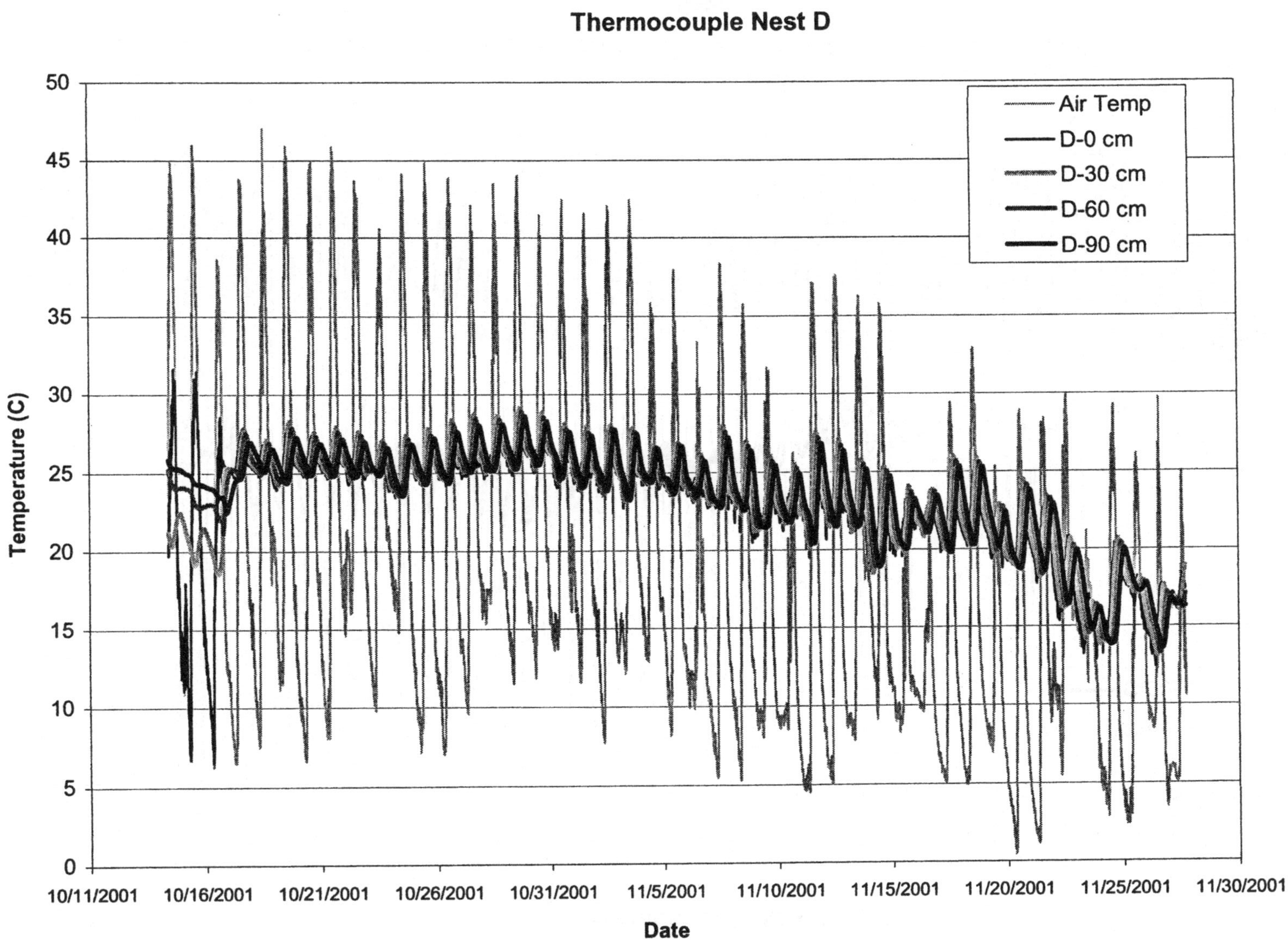

Figure G.2 (continued) Thermocouple data 10/14/01 through 11/27/02

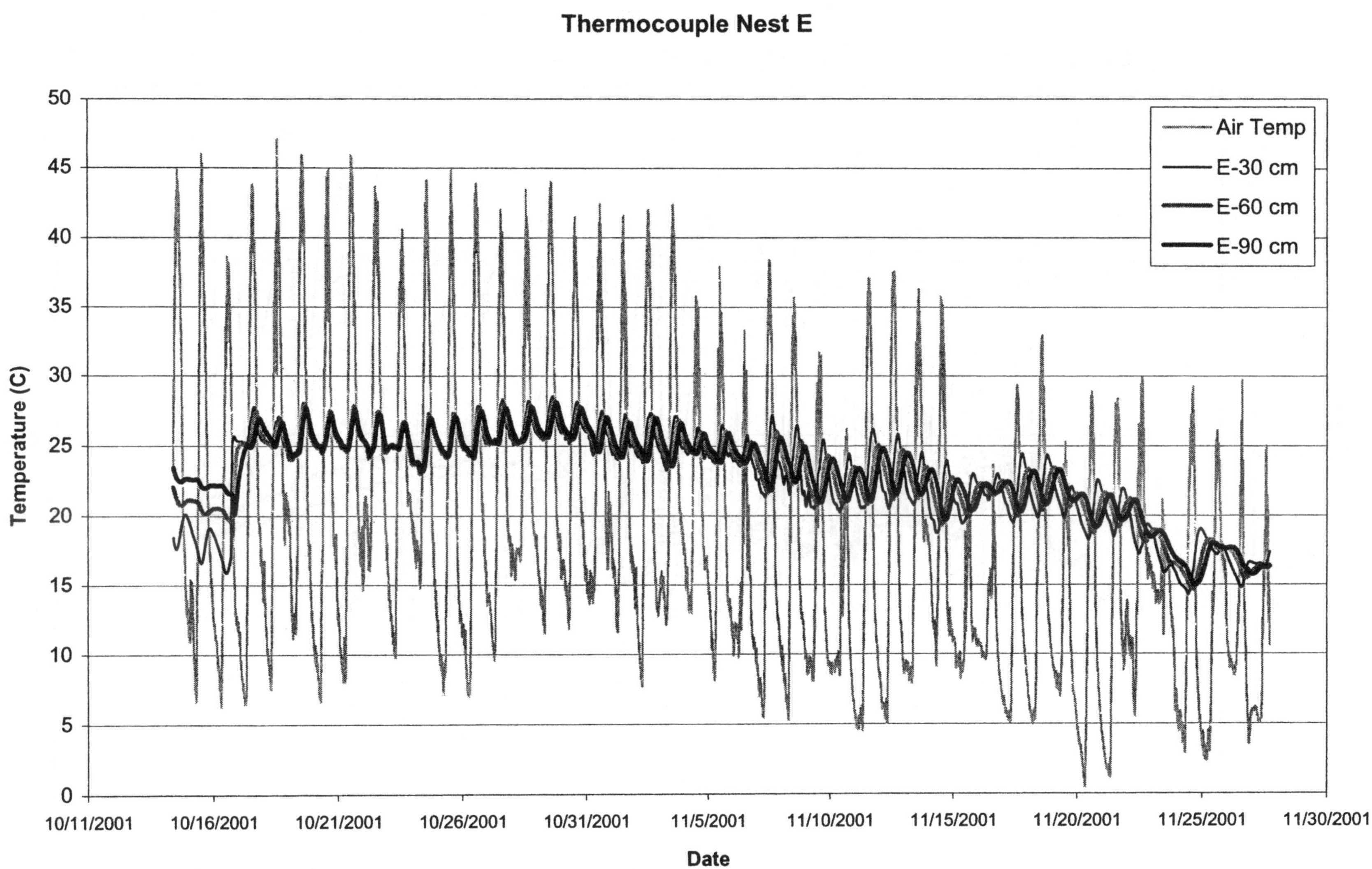

Figure G.2 (continued) Thermocouple data 10/14/01 through 11/27/02

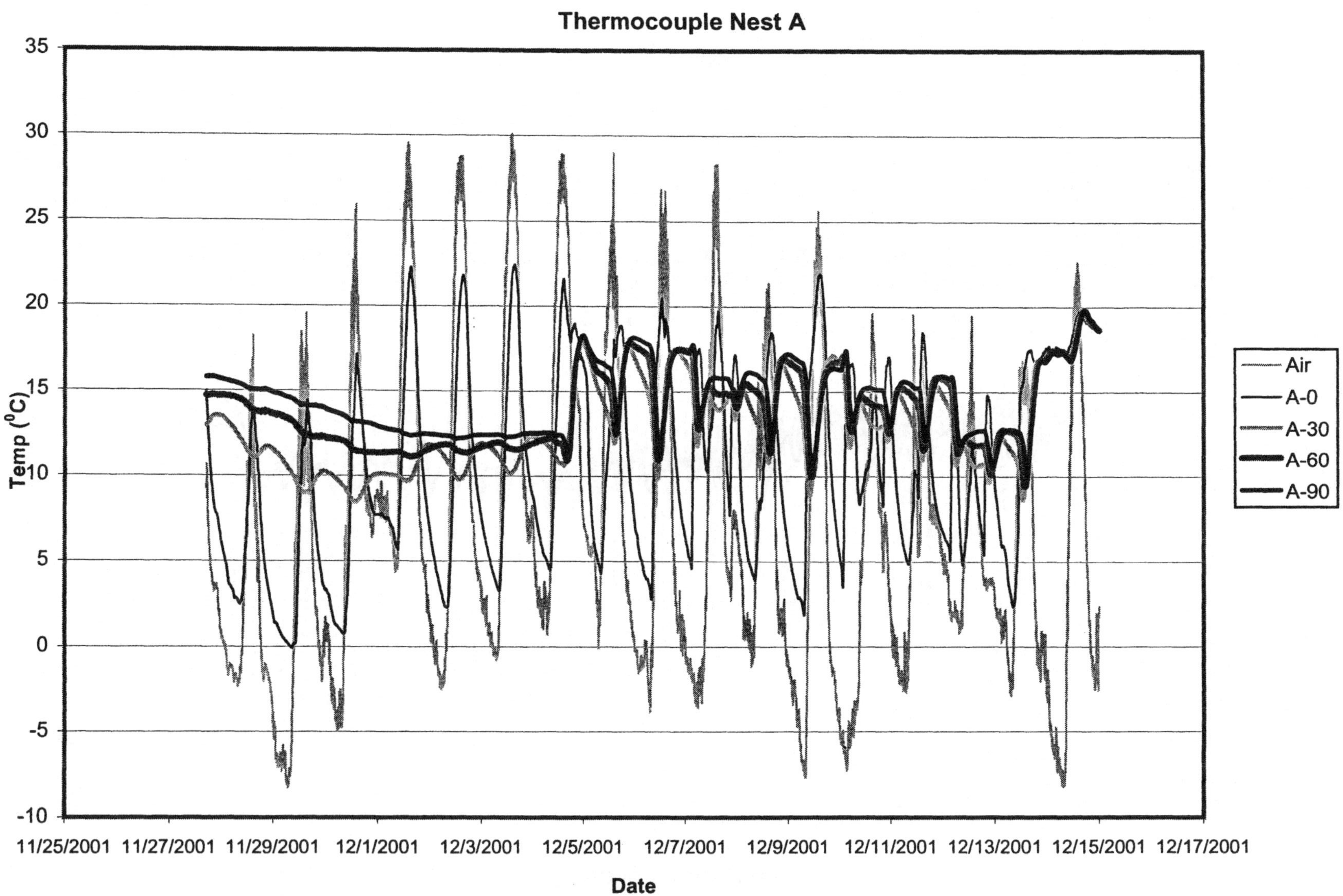

Figure G.3 Thermocouple data 11/27/01 through 1/28/02

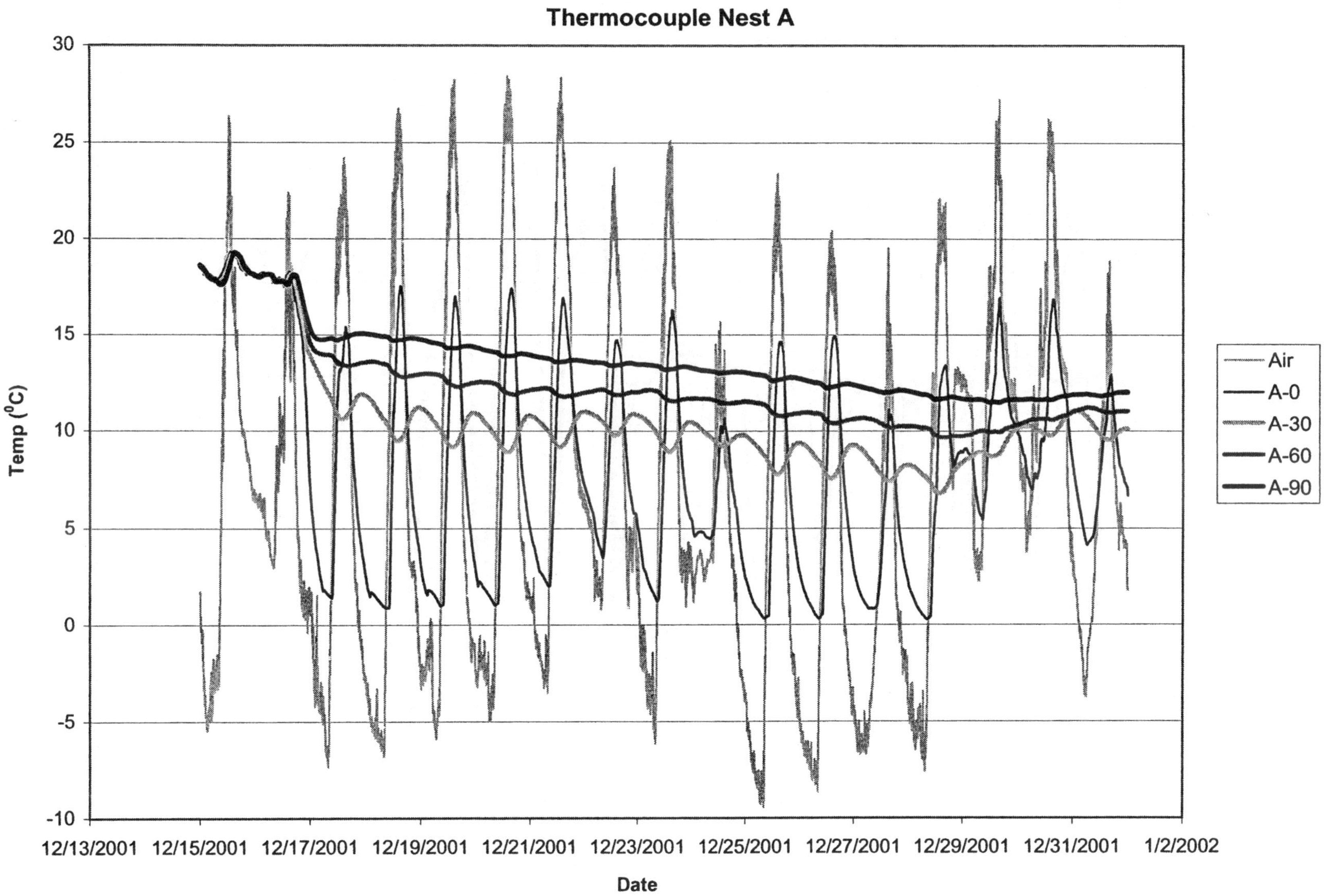

Figure G.3 (continued) Thermocouple data 11/27/01 through 1/28/02

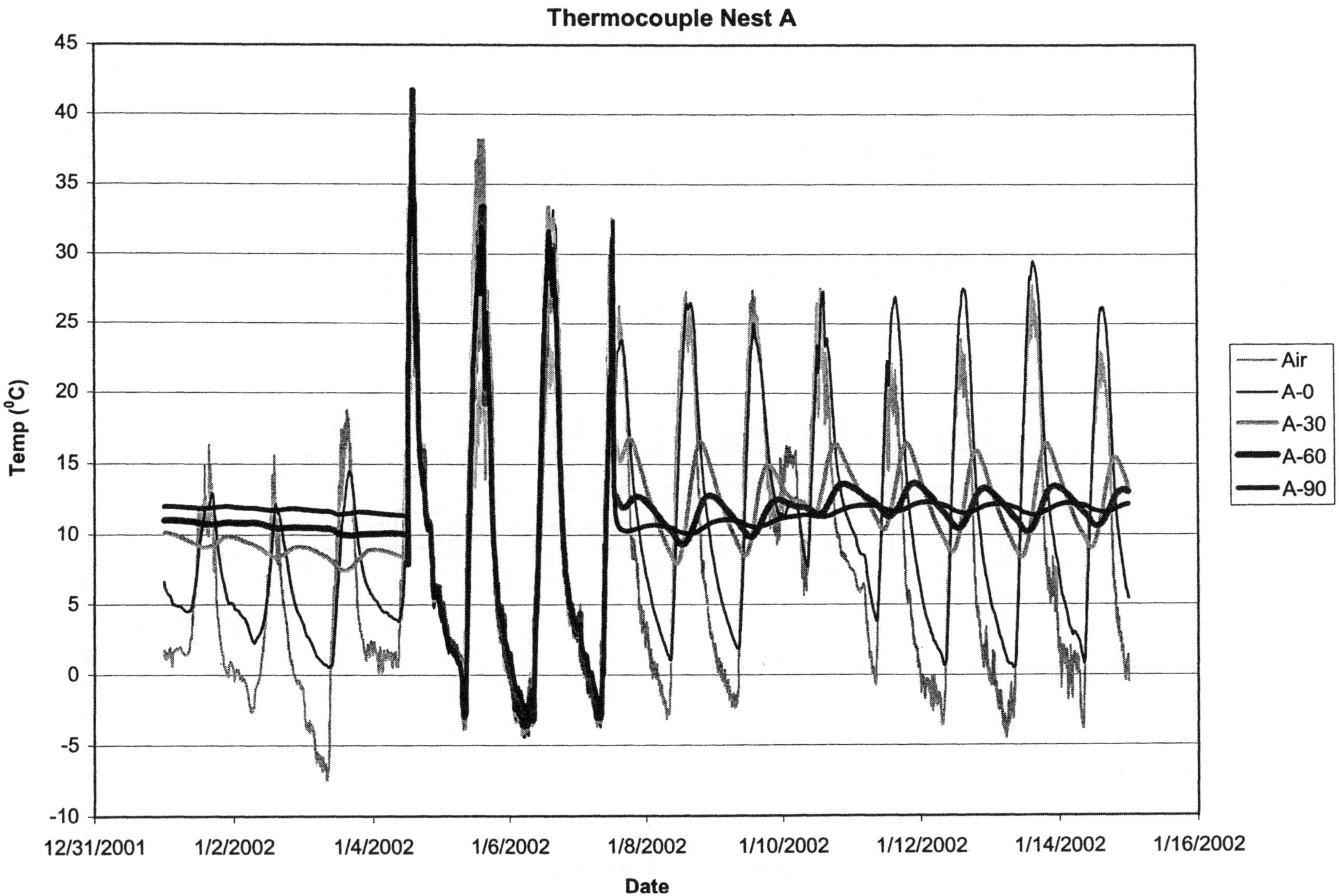

Figure G.3 (continued) Thermocouple data 11/27/01 through 1/28/02

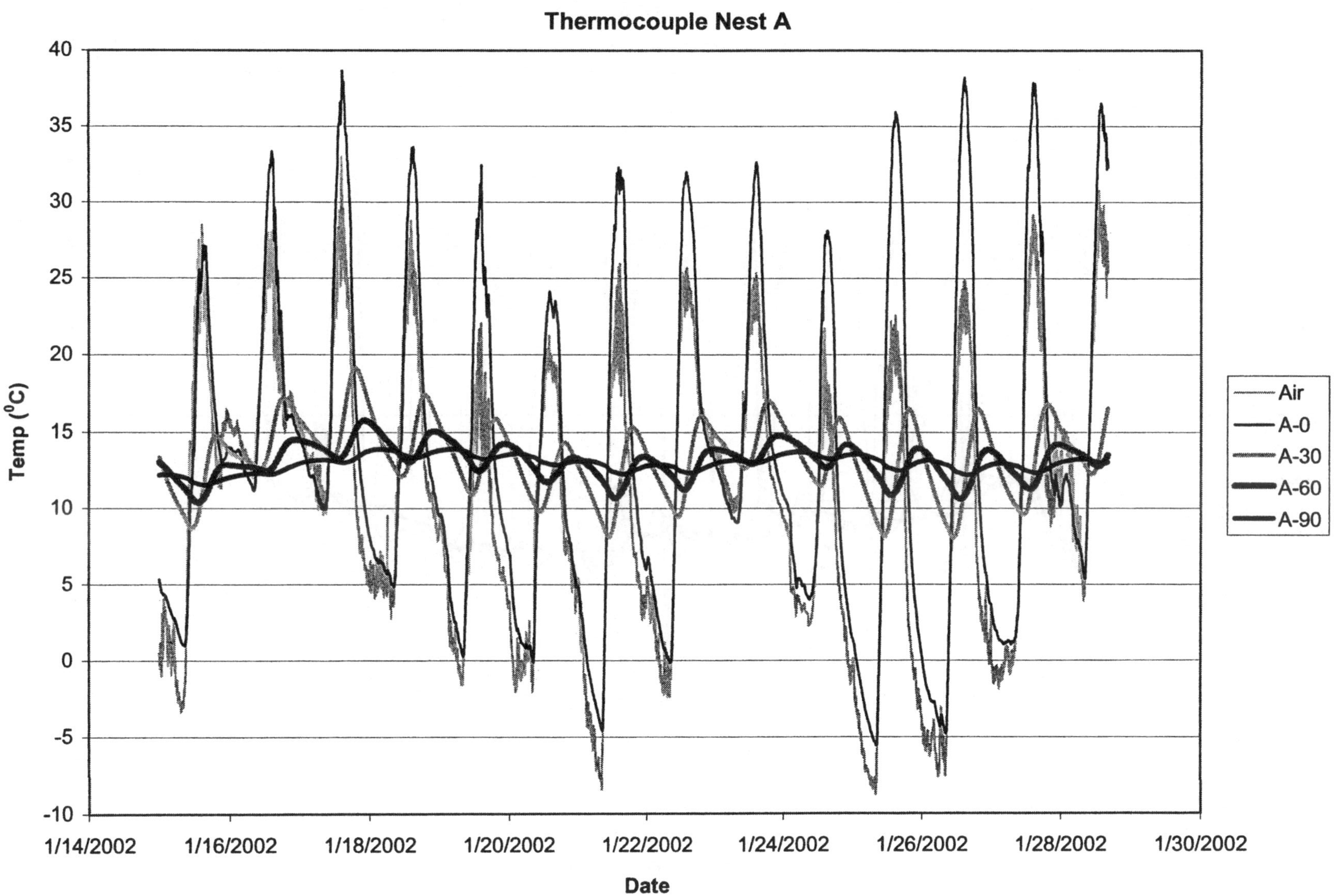

Figure G.3 (continued) Thermocouple data 11/27/01 through 1/28/02

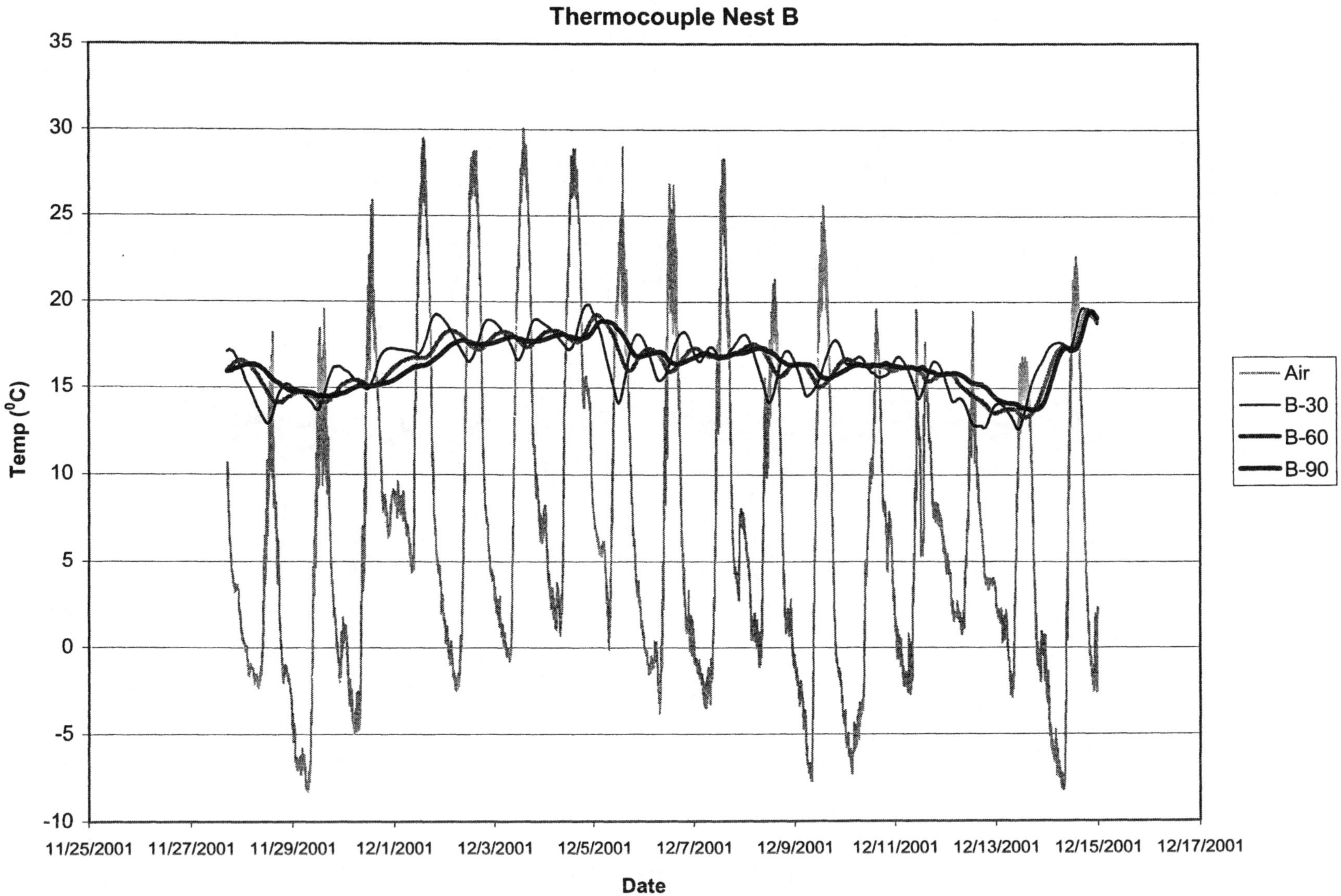

Figure G.3 (continued) Thermocouple data 11/27/01 through 1/28/02

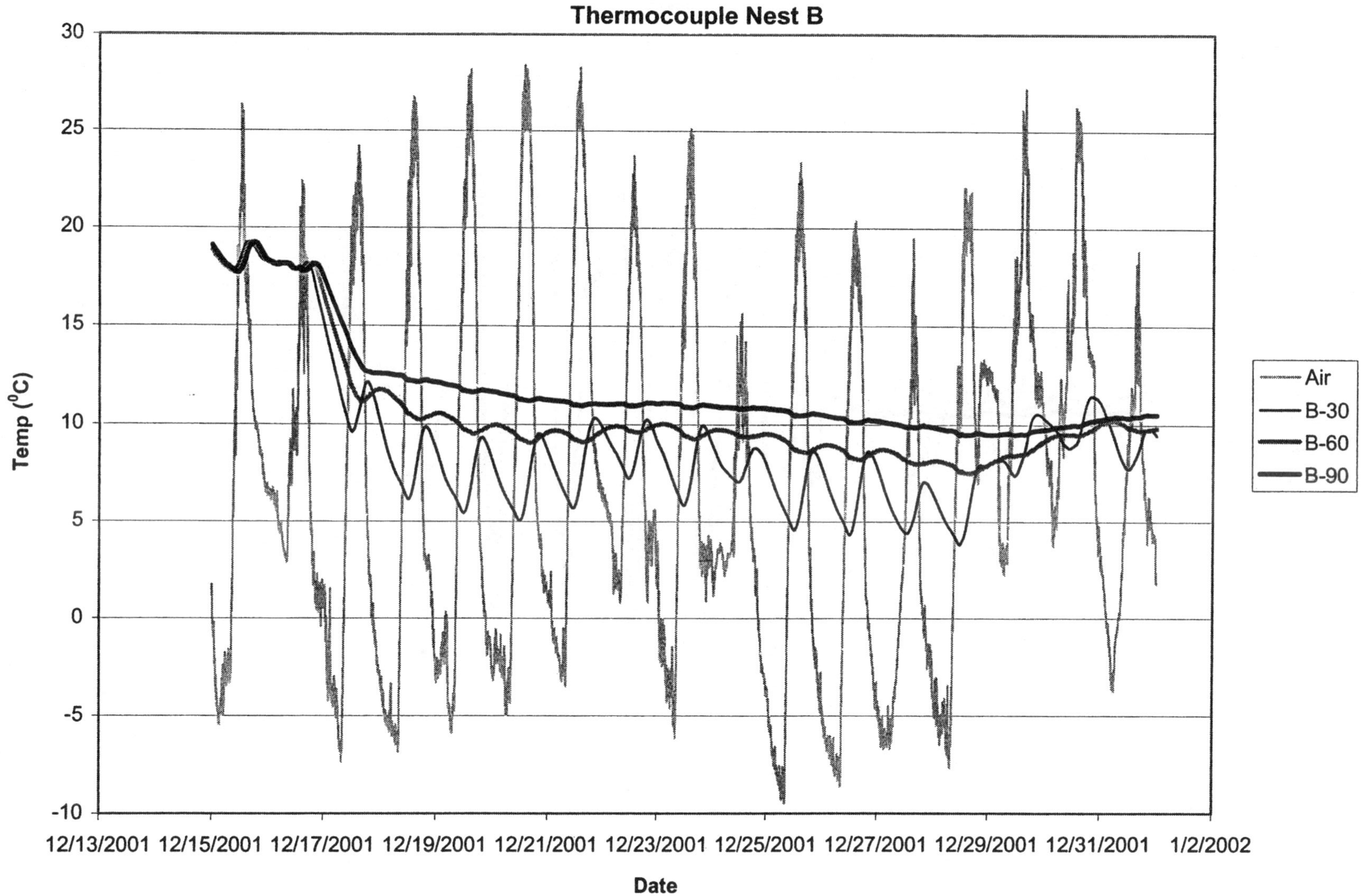

Figure G.3 (continued) Thermocouple data 11/27/01 through 1/28/02

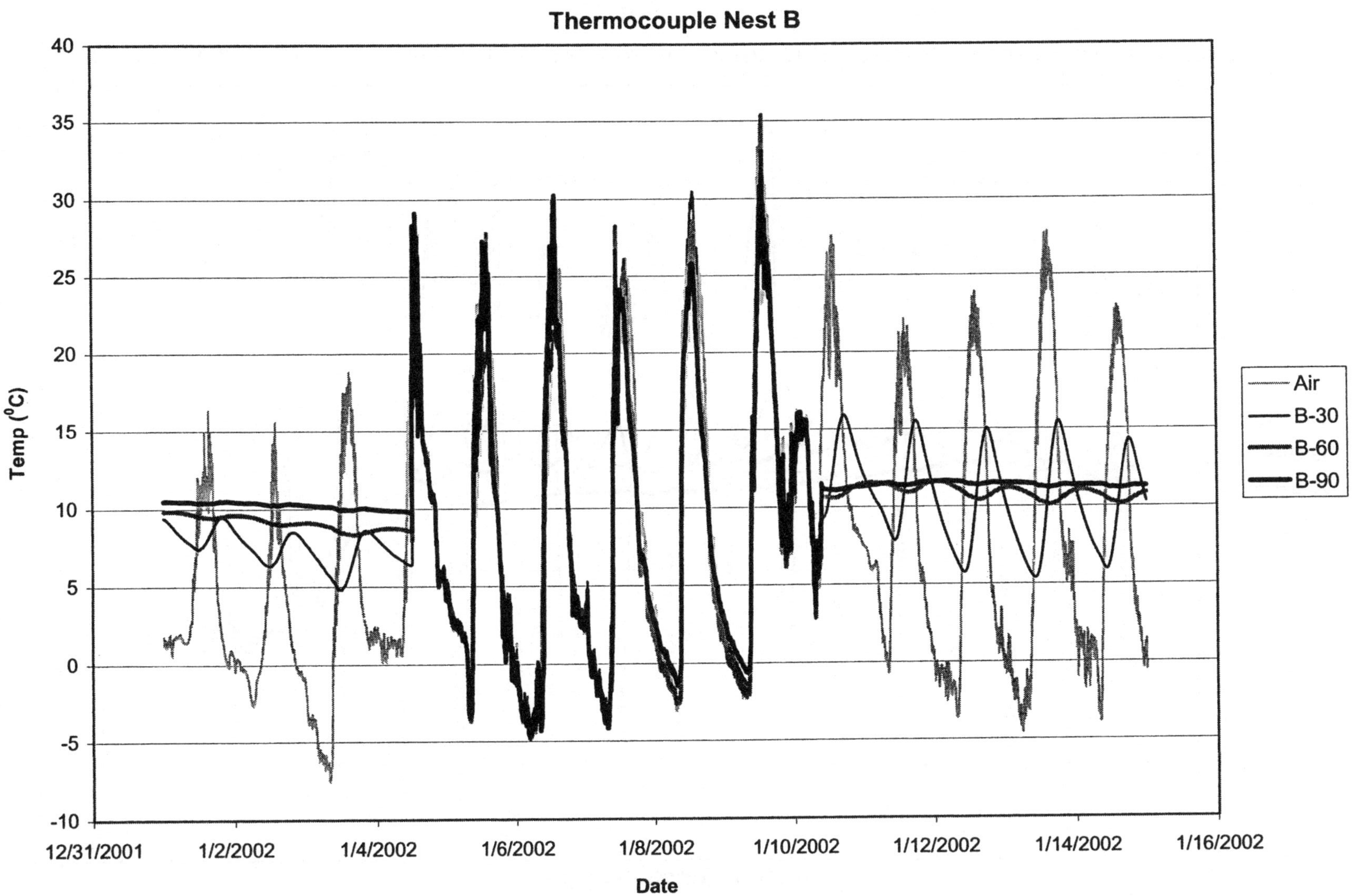

Figure G.3 (continued) Thermocouple data 11/27/01 through 1/28/02

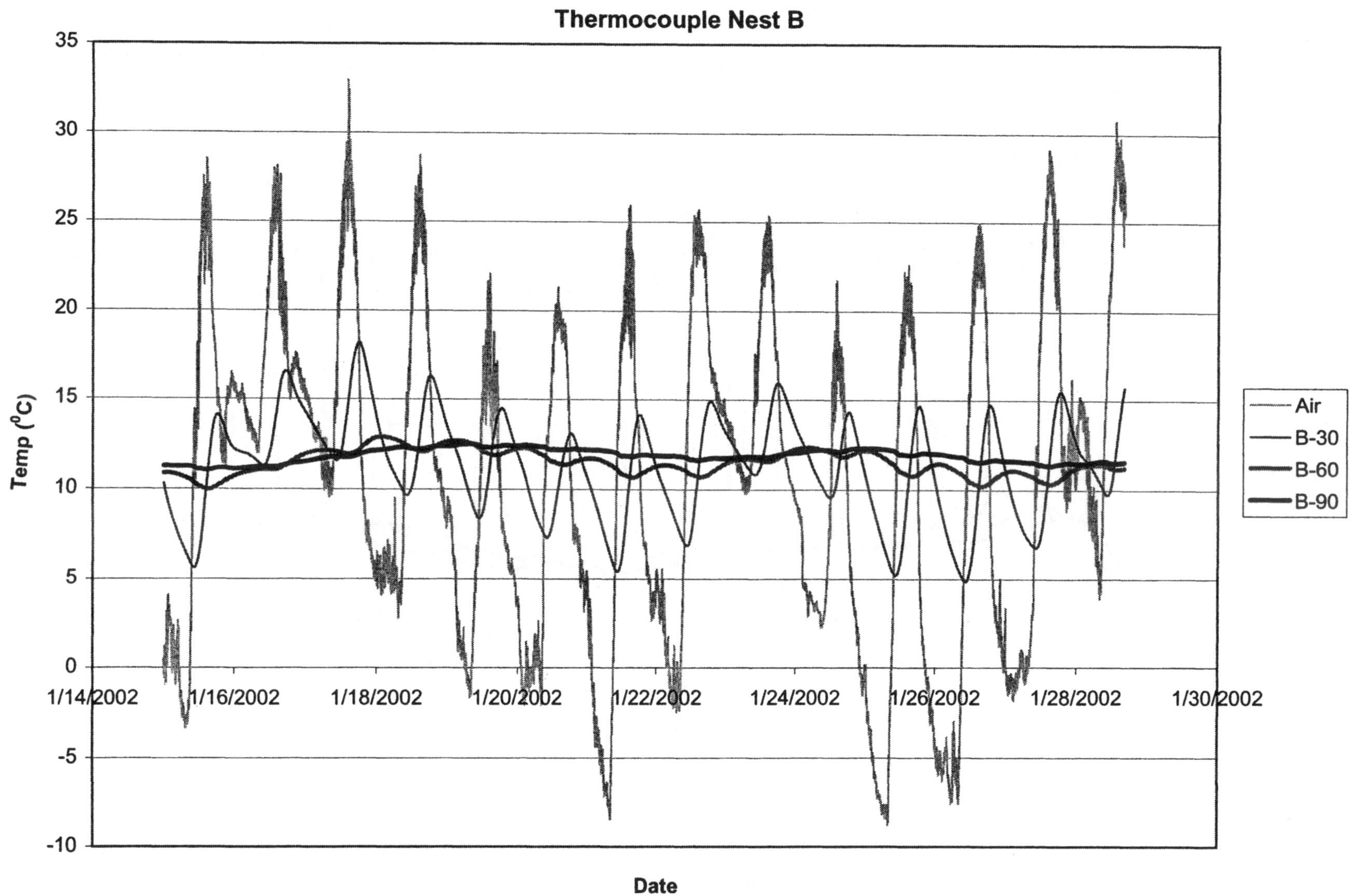

Figure G.3 (continued) Thermocouple data 11/27/01 through 1/28/02

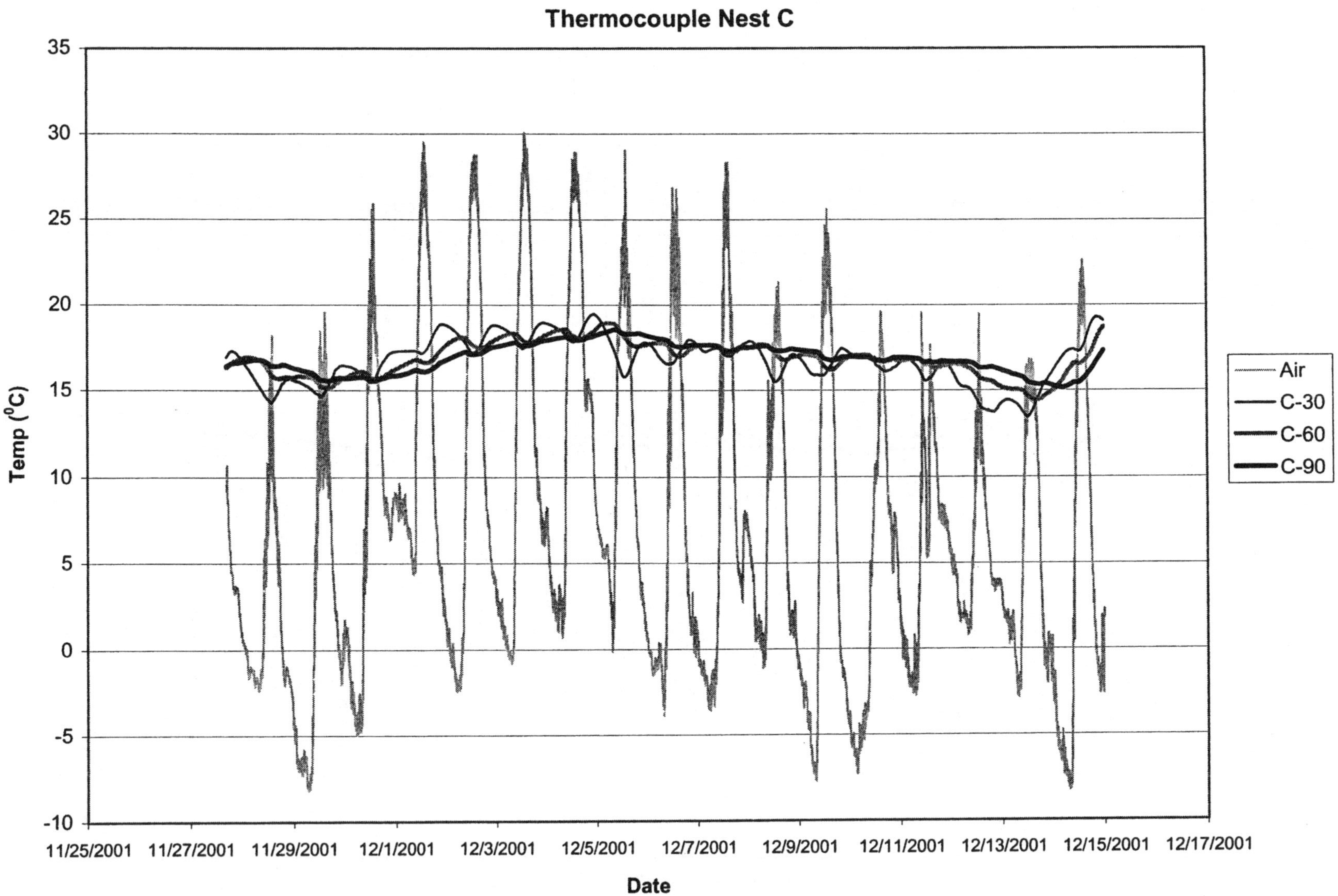

Figure G.3 (continued) Thermocouple data 11/27/01 through 1/28/02

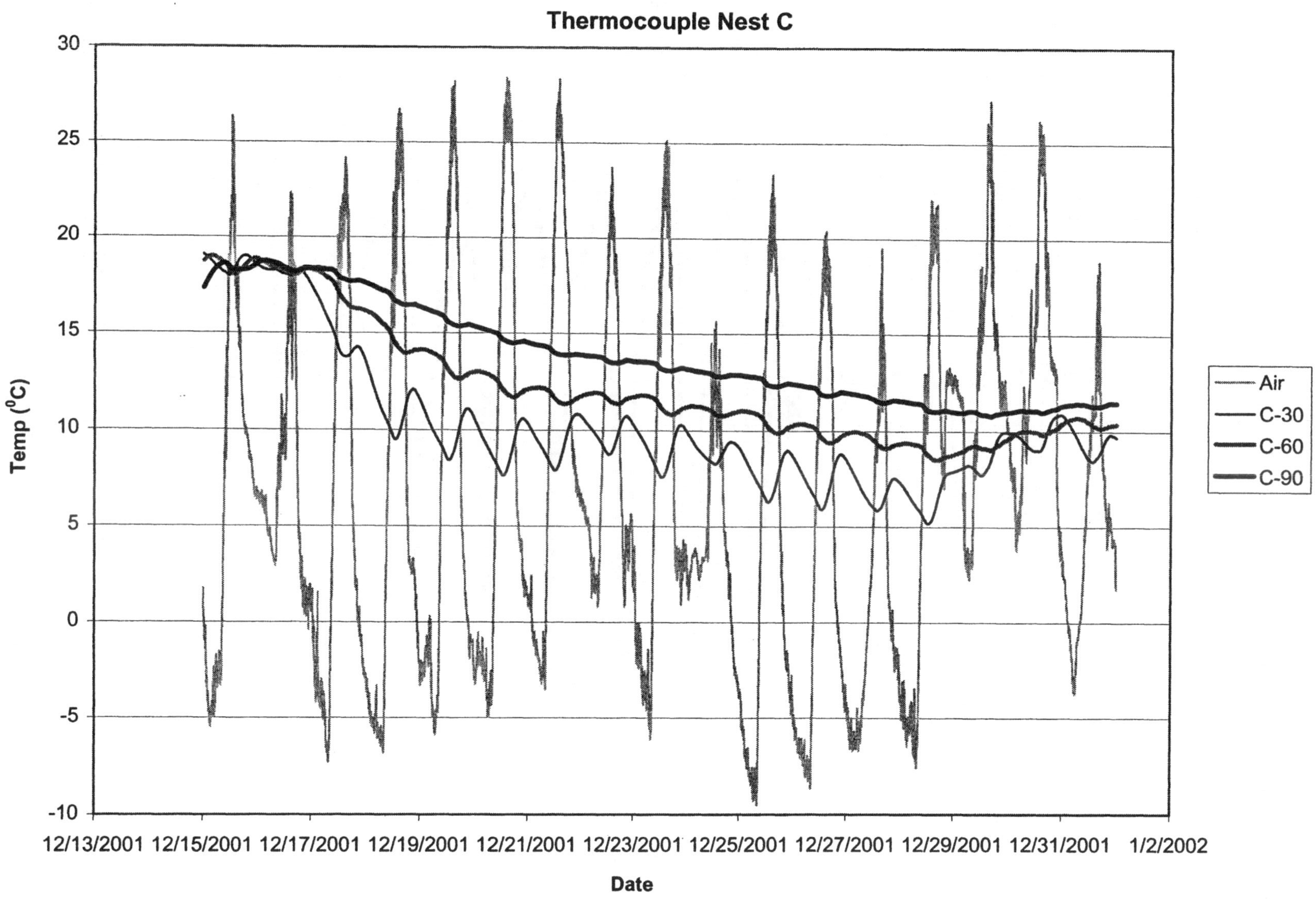

Figure G.3 (continued) Thermocouple data 11/27/01 through 1/28/02

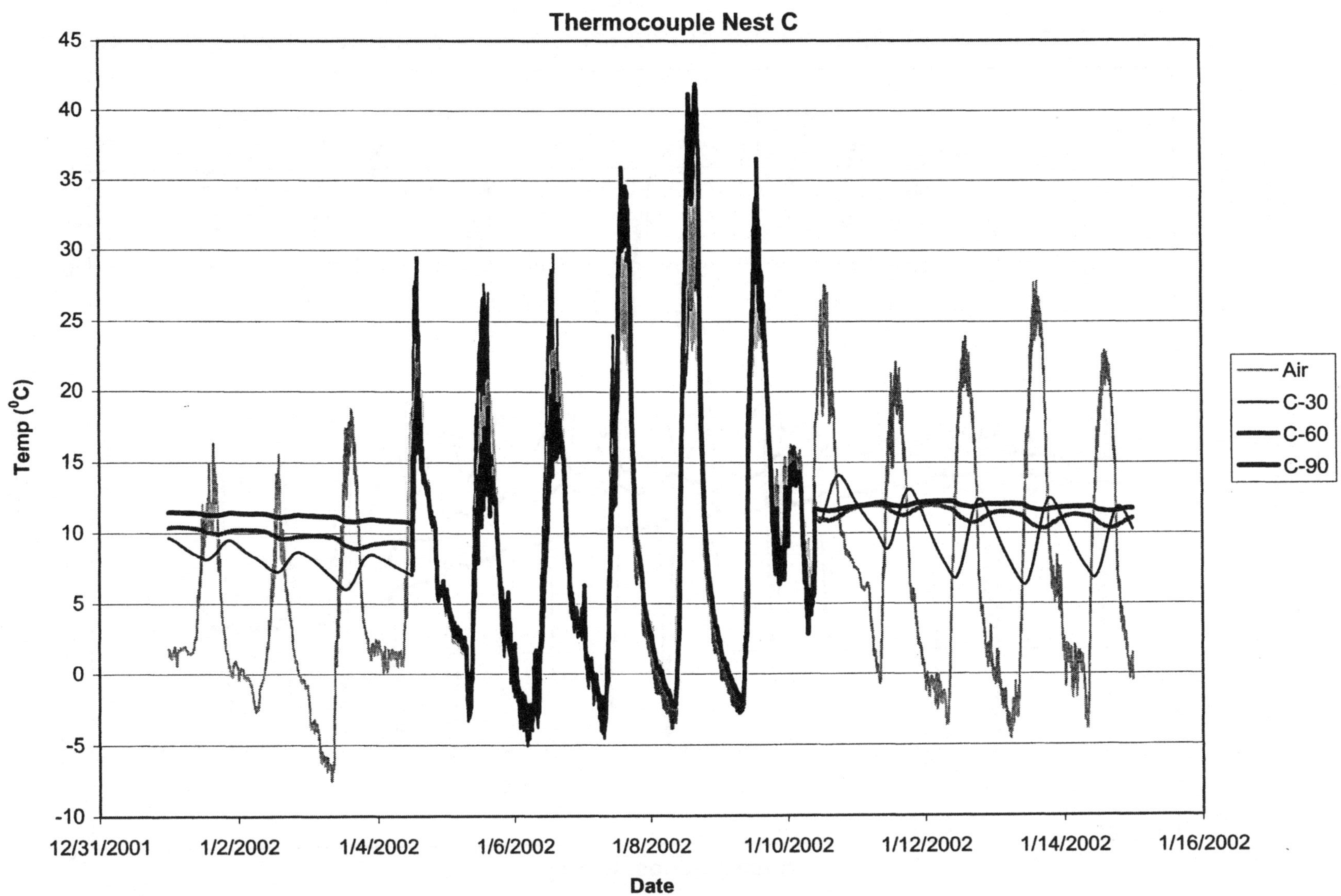

Figure G.3 (continued) Thermocouple data 11/27/01 through 1/28/02

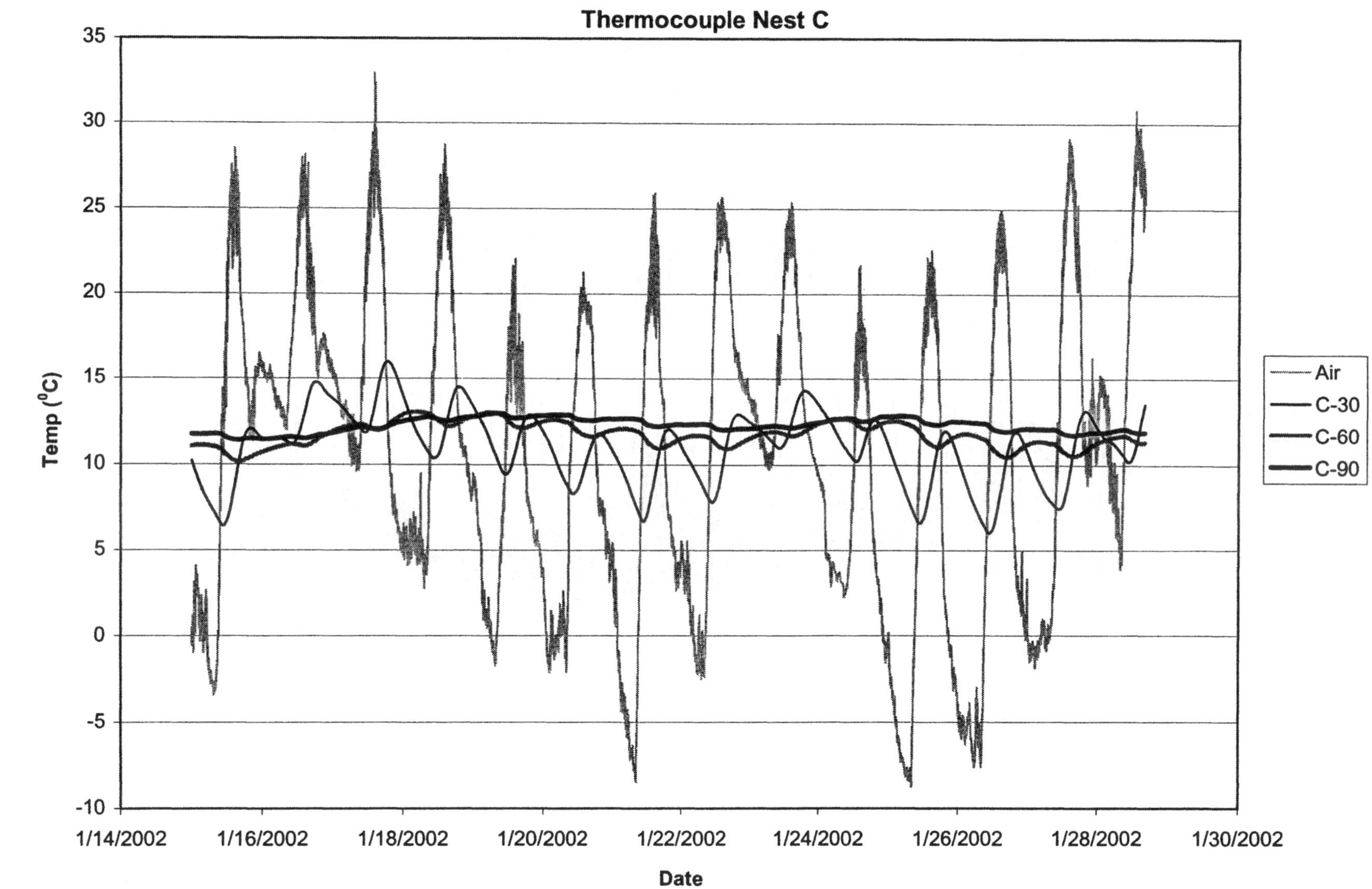

Figure G.3 (continued) Thermocouple data 11/27/01 through 1/28/02

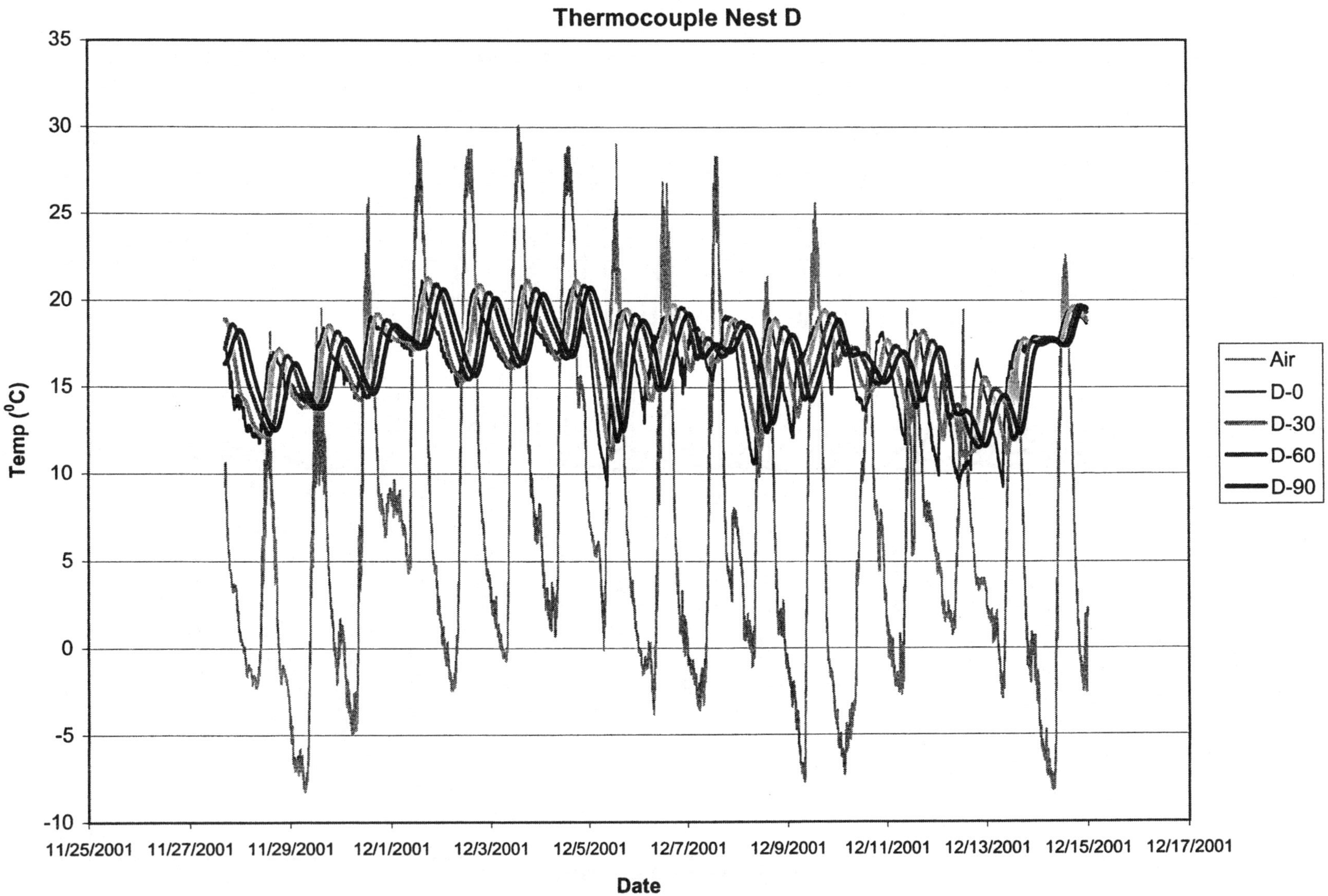

Figure G.3 (continued) Thermocouple data 11/27/01 through 1/28/02

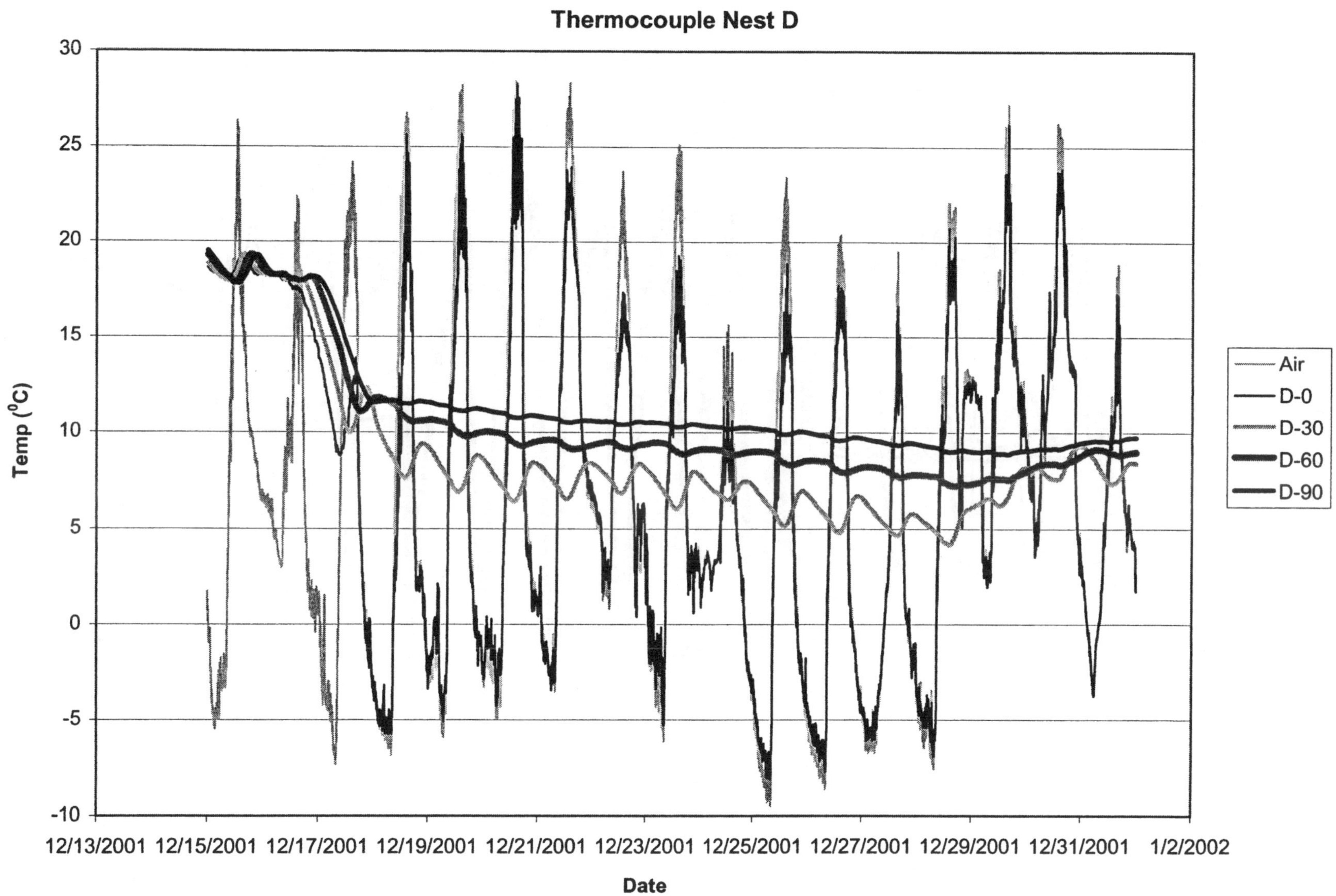

Figure G.3 (continued) Thermocouple data 11/27/01 through 1/28/02

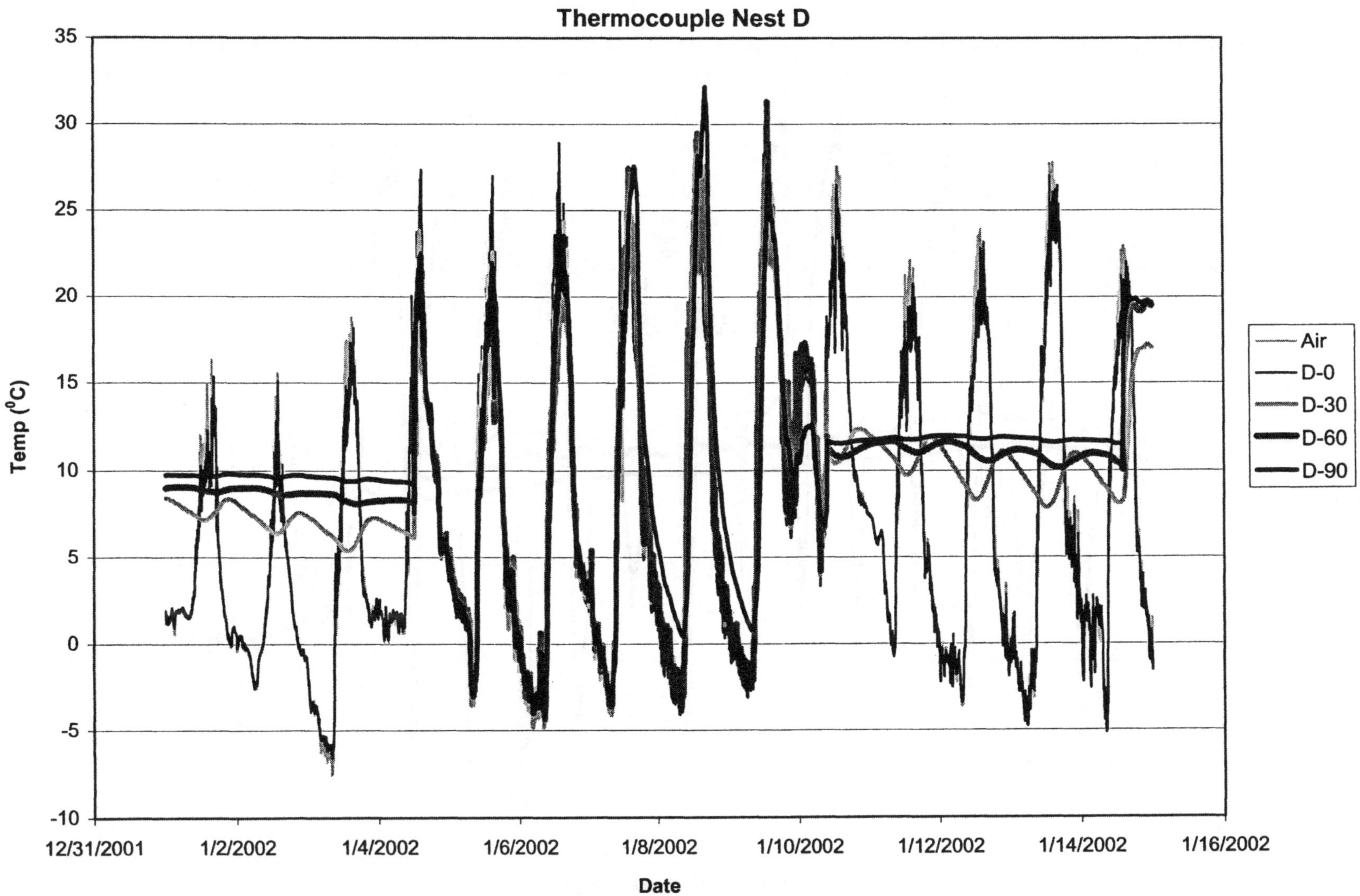

Figure G.3 (continued) Thermocouple data 11/27/01 through 1/28/02

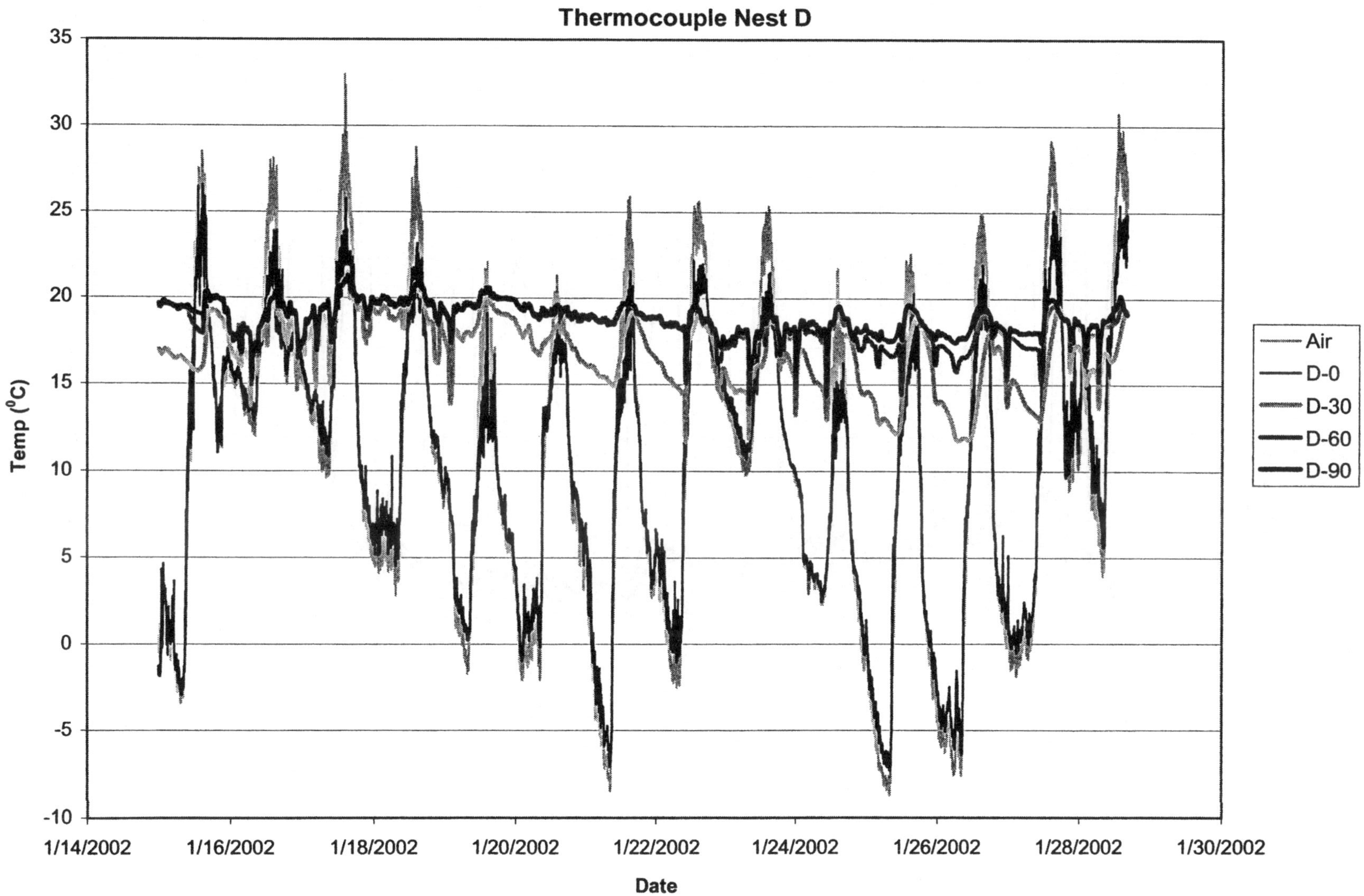

Figure G.3 (continued) Thermocouple data 11/27/01 through 1/28/02

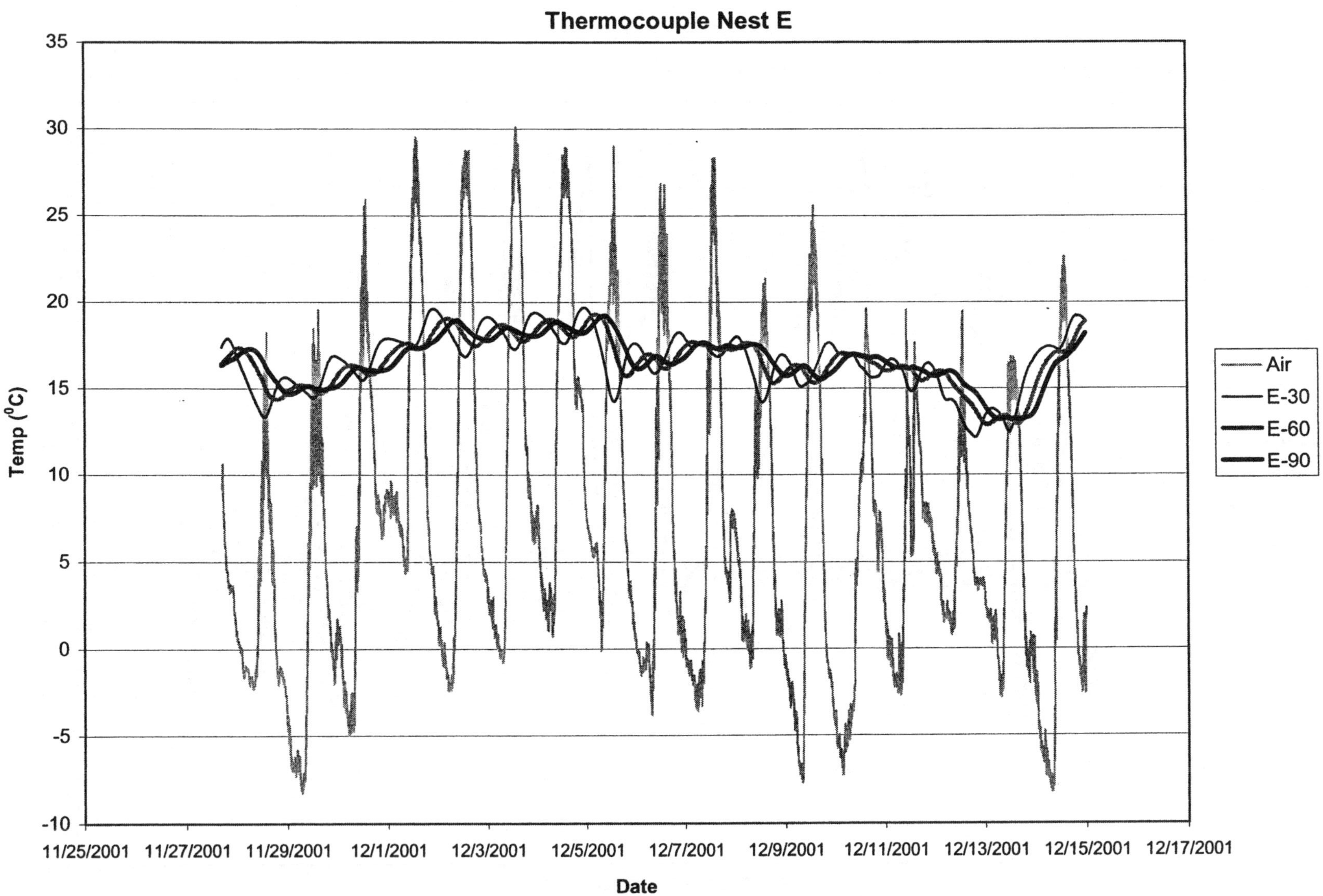

Figure G.3 (continued) Thermocouple data 11/27/01 through 1/28/02

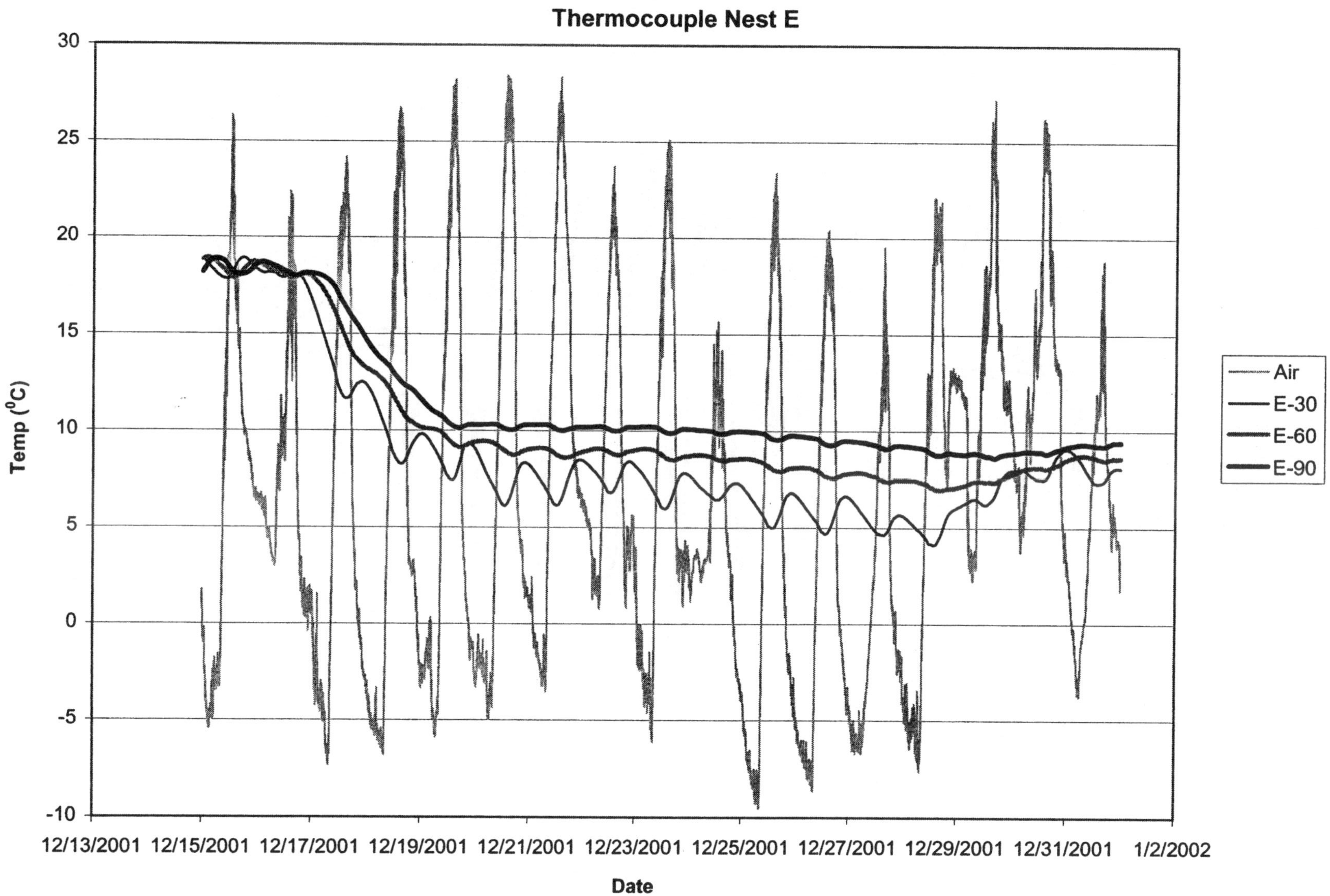

Figure G.3 (continued) Thermocouple data 11/27/01 through 1/28/02

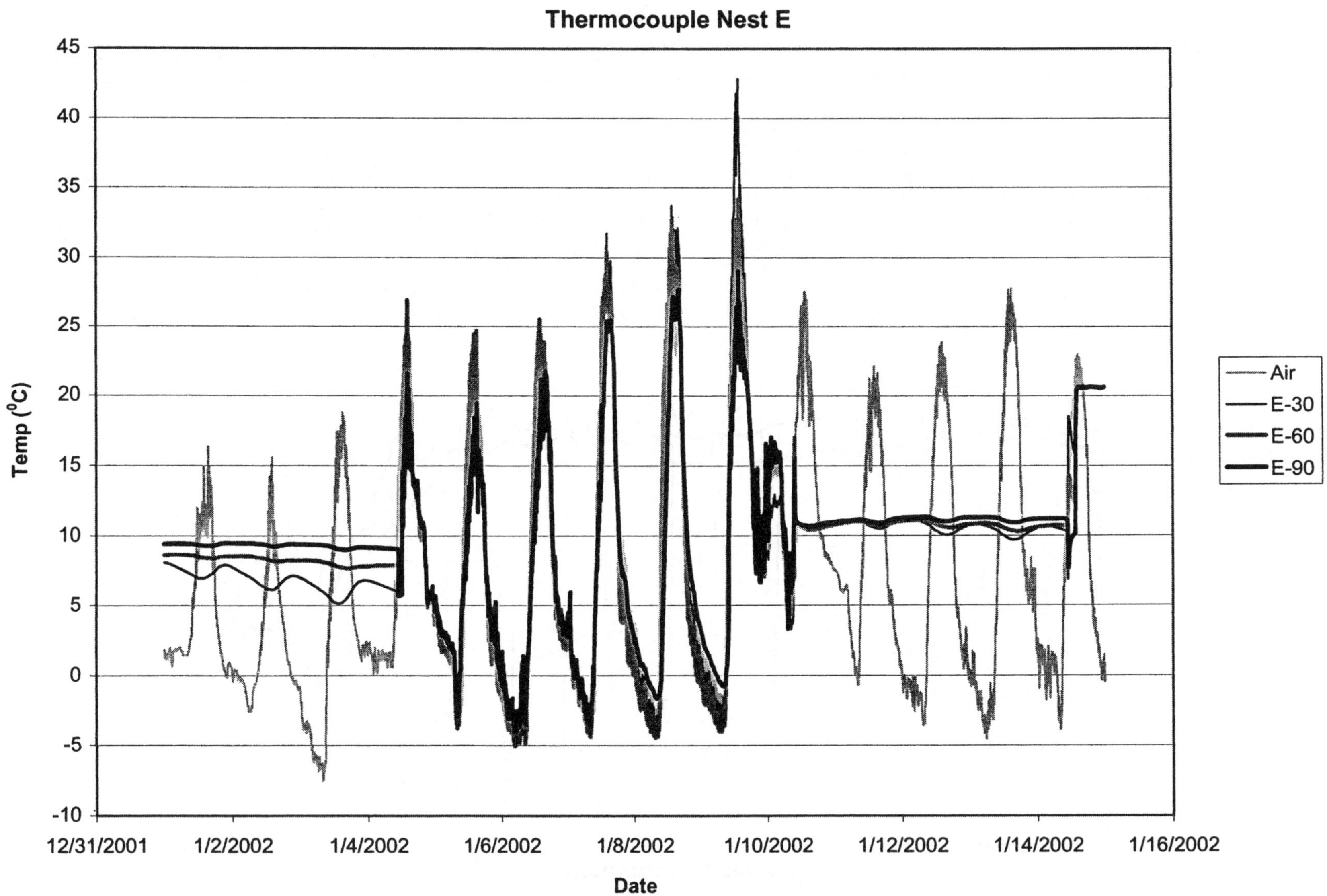

Figure G.3 (continued) Thermocouple data 11/27/01 through 1/28/02

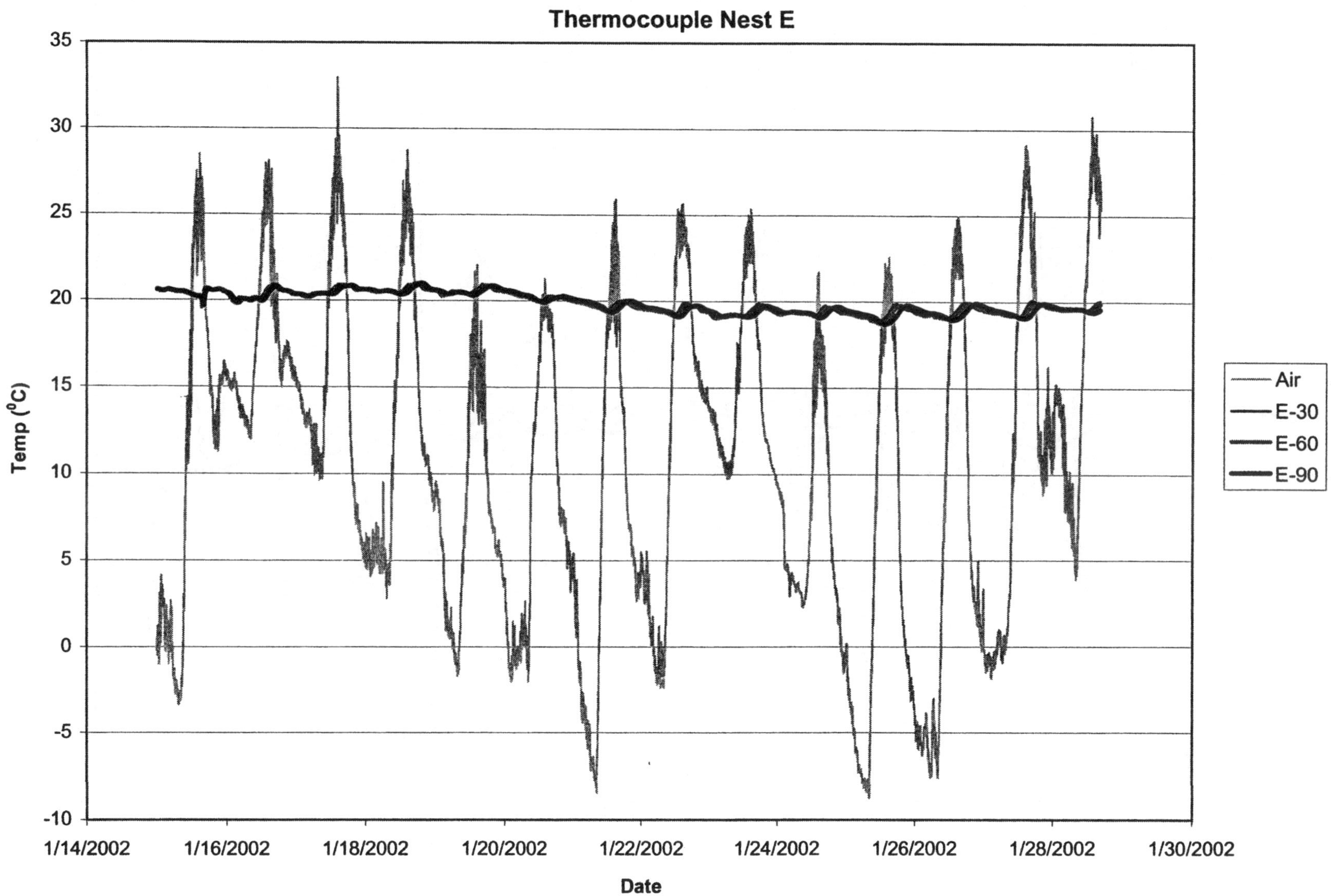

Figure G.3 (continued) Thermocouple data 11/27/01 through 1/28/02

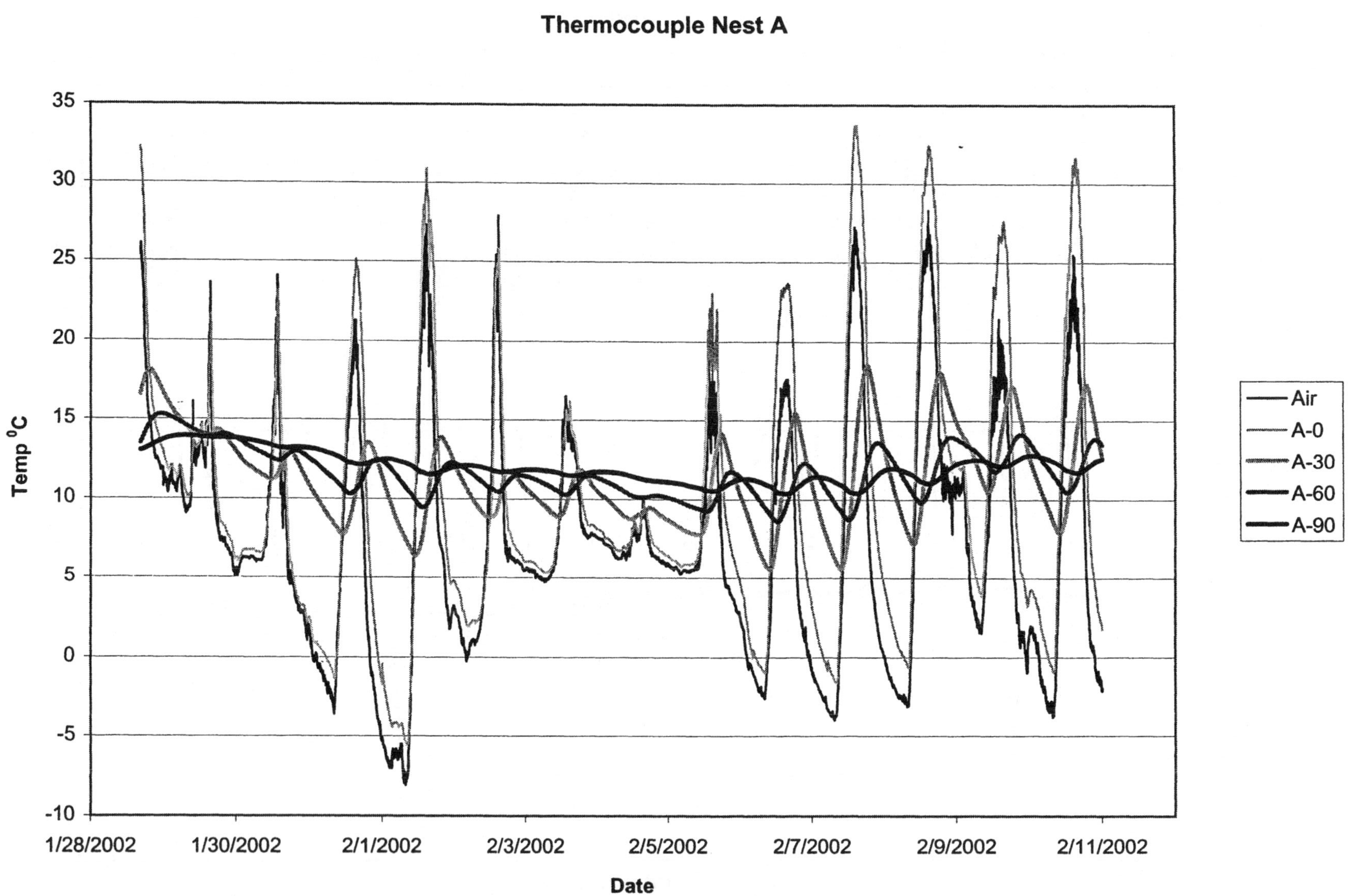

Figure G.4 Thermocouple data 1/28/02 through 3/22/02

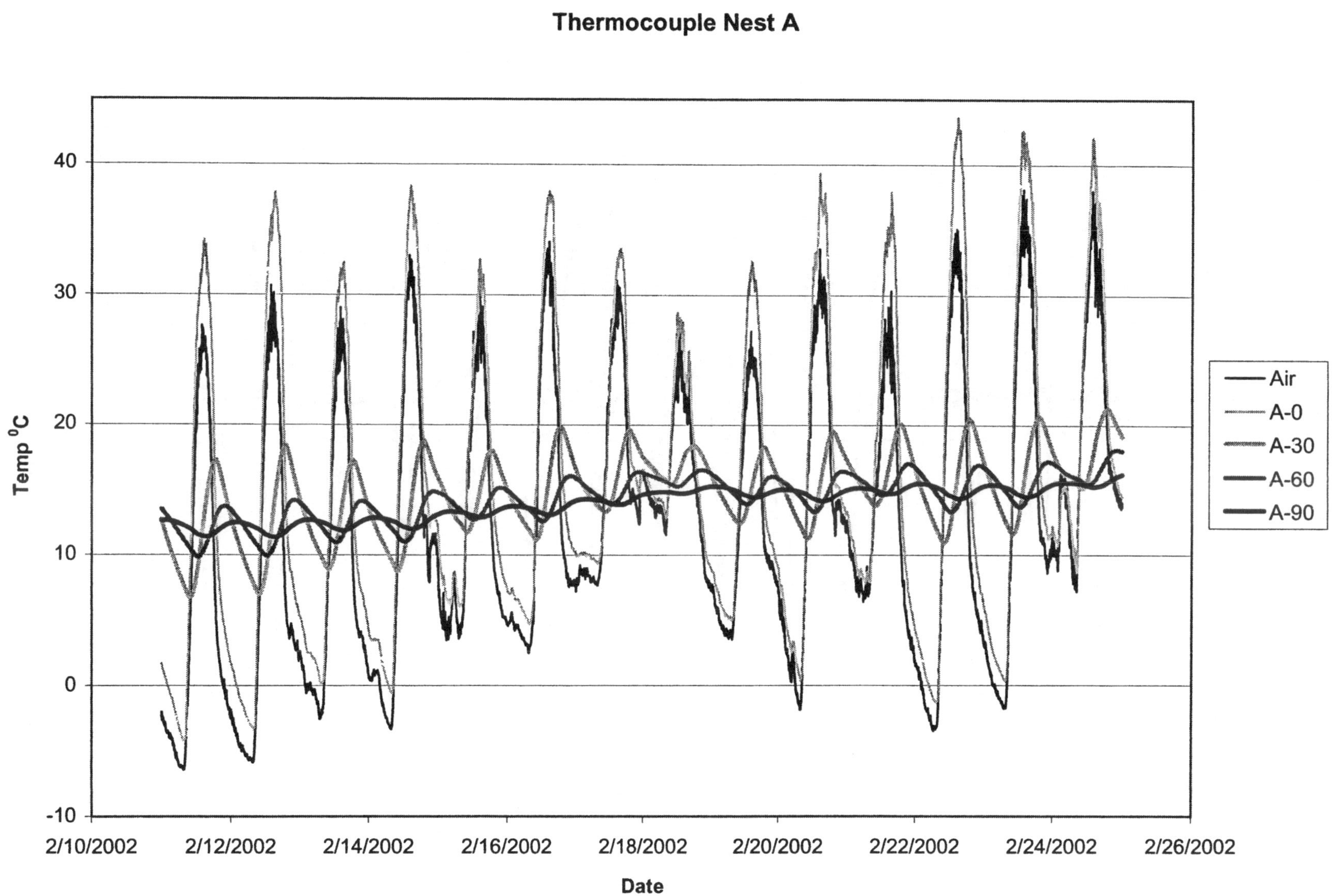

Figure G.4 (continued) Thermocouple data 1/28/02 through 3/22/02

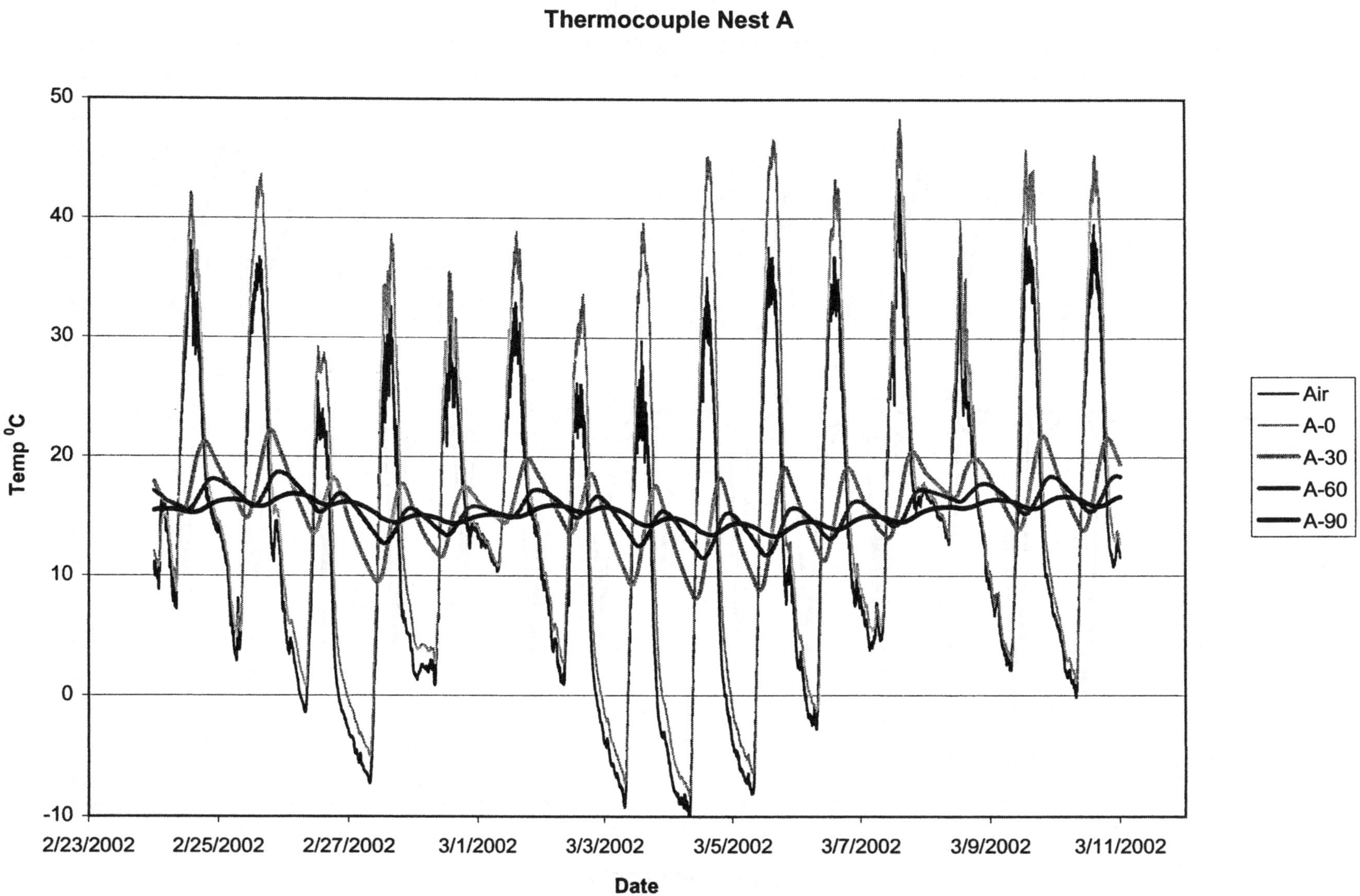

Figure G.4 (continued) Thermocouple data 1/28/02 through 3/22/02

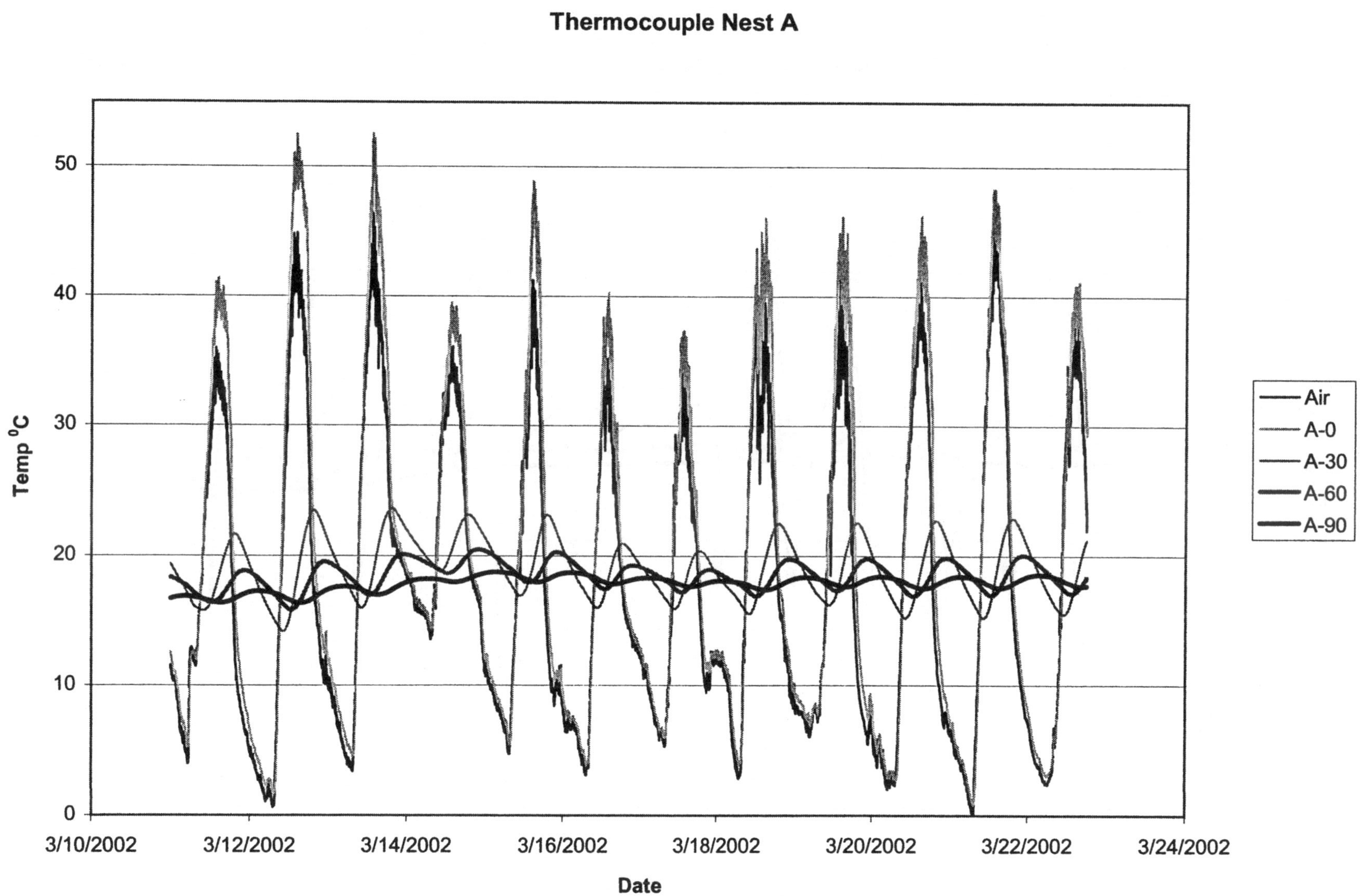

Figure G.4 (continued) Thermocouple data 1/28/02 through 3/22/02

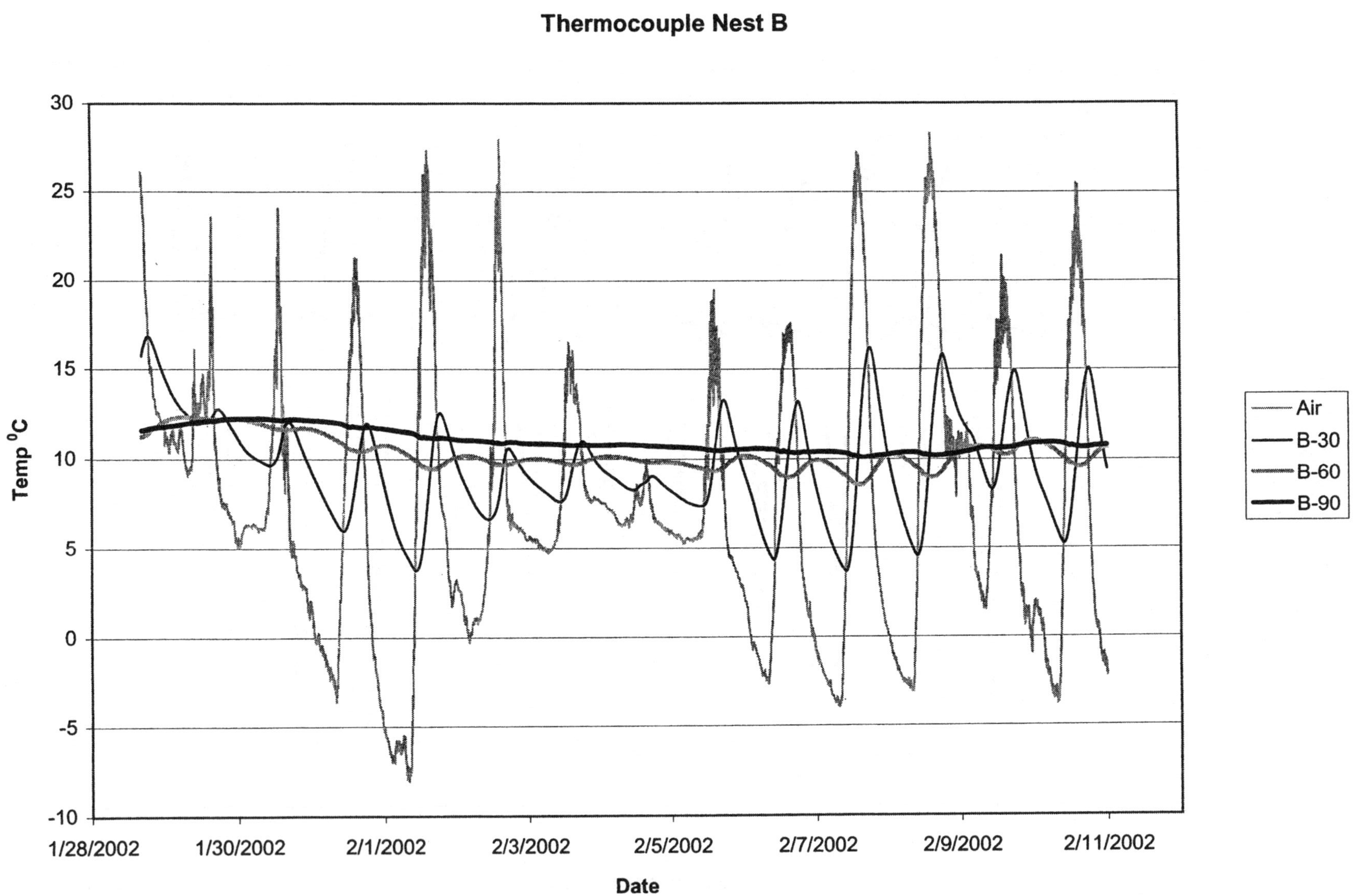

Figure G.4 (continued) Thermocouple data 1/28/02 through 3/22/02

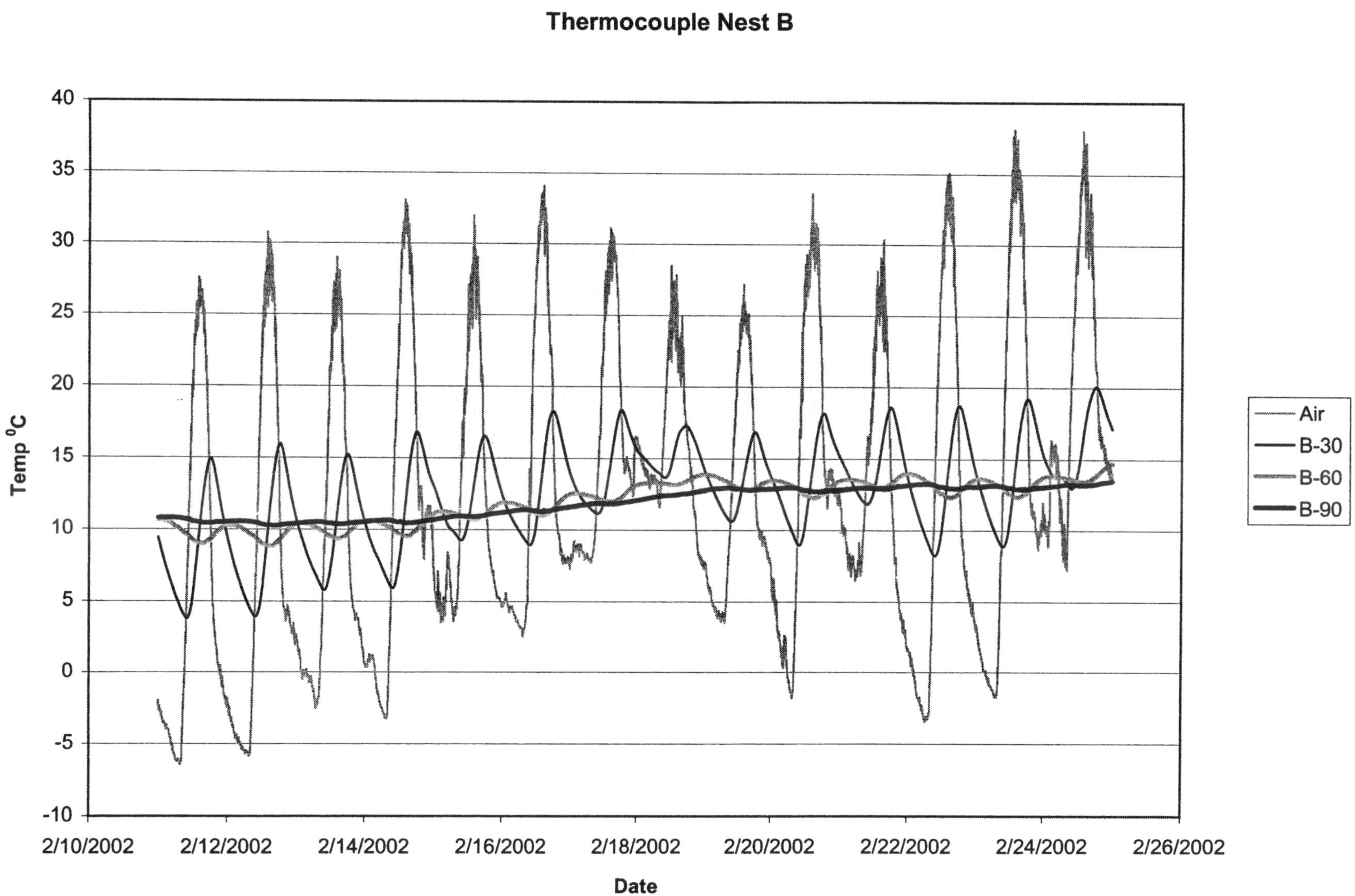

Figure G.4 (continued) Thermocouple data 1/28/02 through 3/22/02

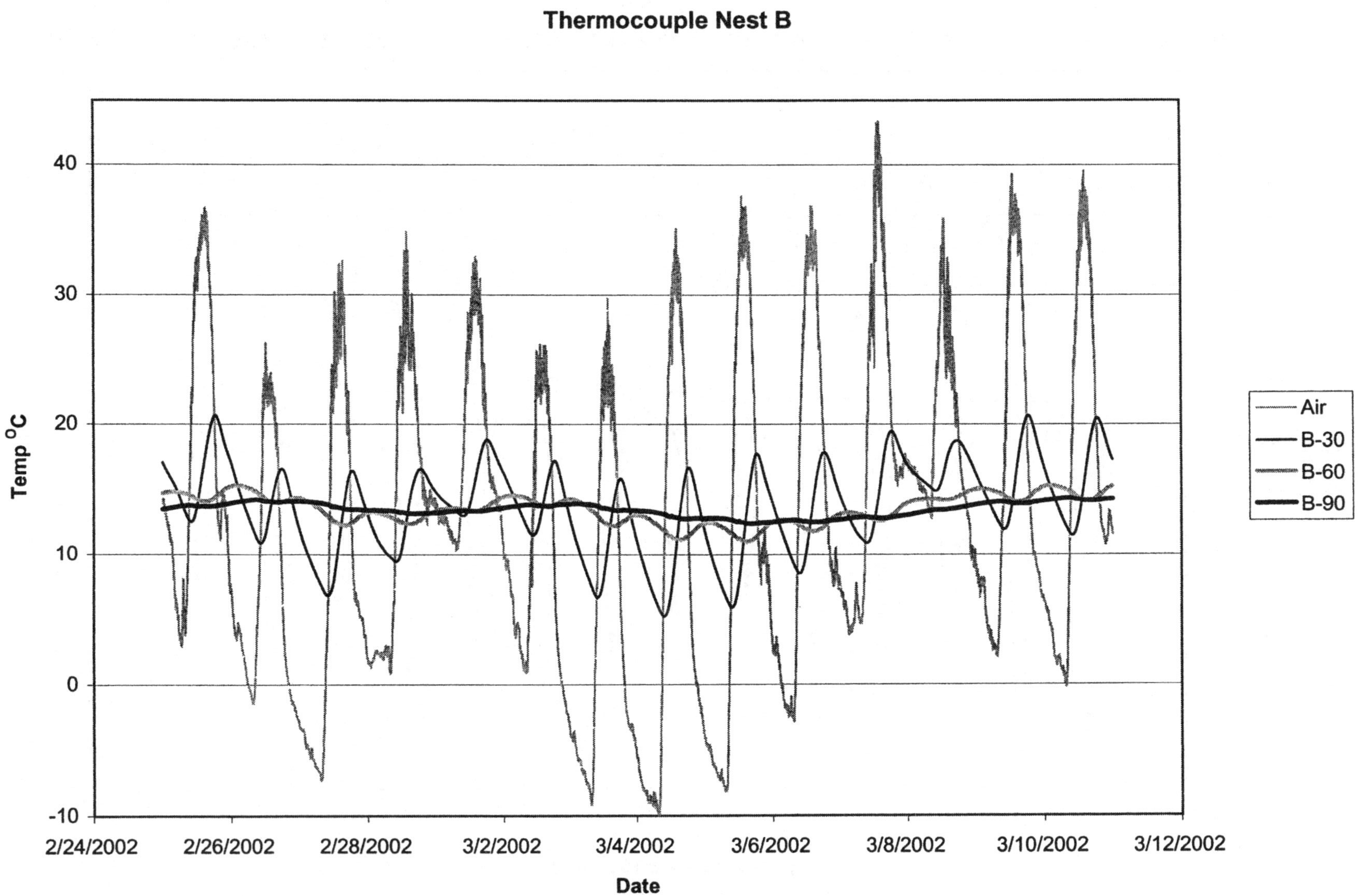

Figure G.4 (continued) Thermocouple data 1/28/02 through 3/22/02

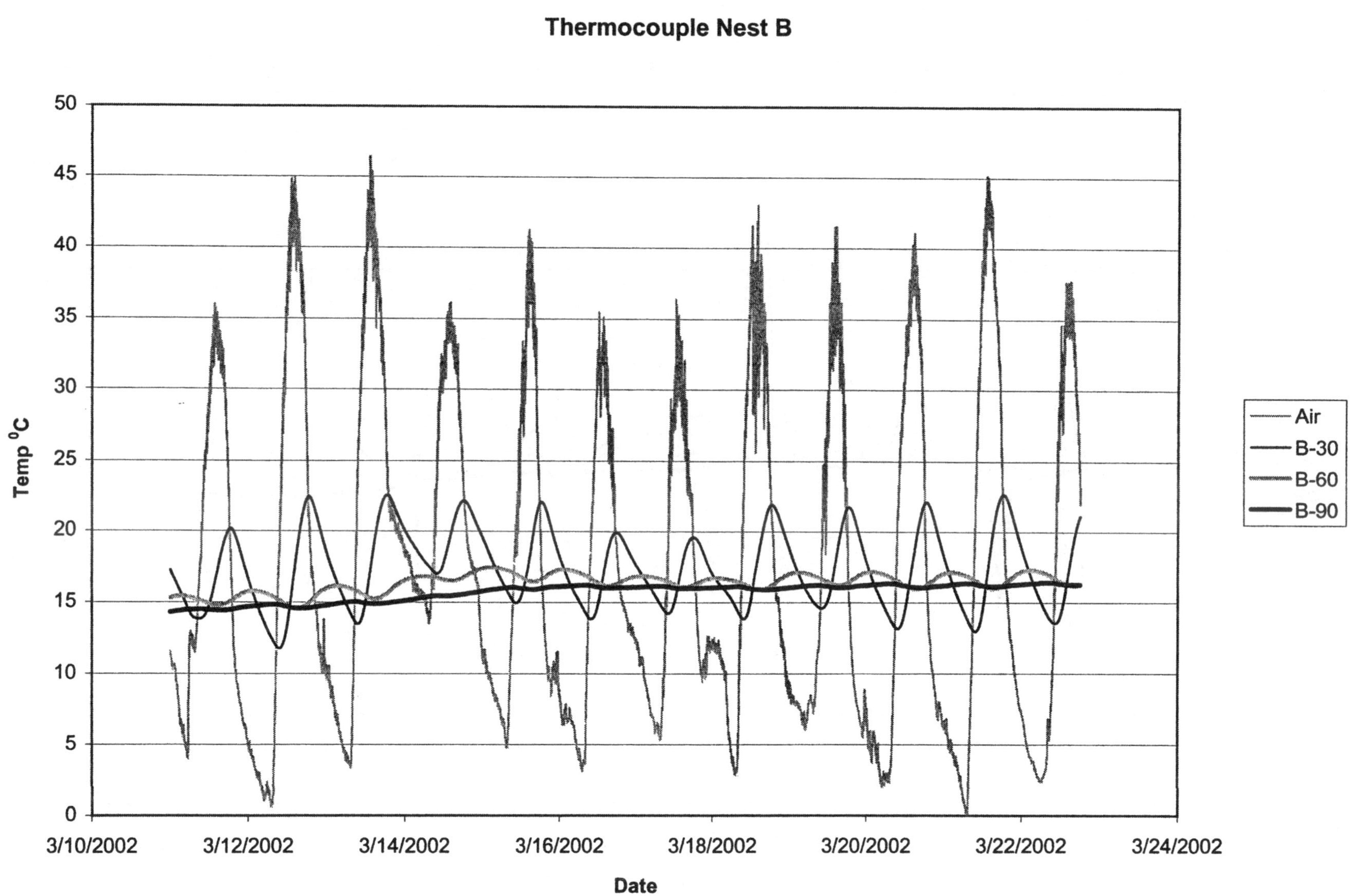

Figure G.4 (continued) Thermocouple data 1/28/02 through 3/22/02

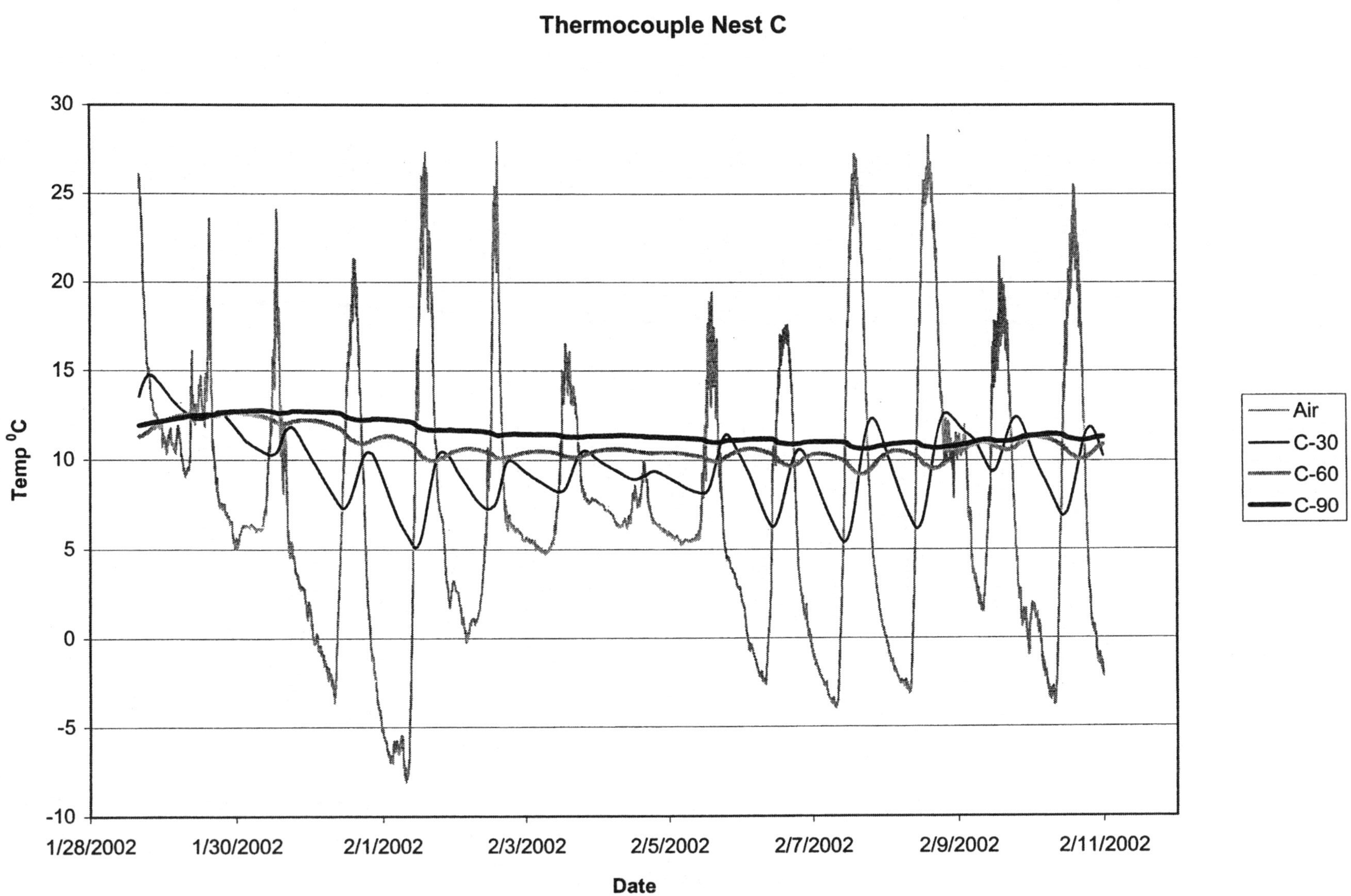

Figure G.4 (continued) Thermocouple data 1/28/02 through 3/22/02

Thermocouple Nest C

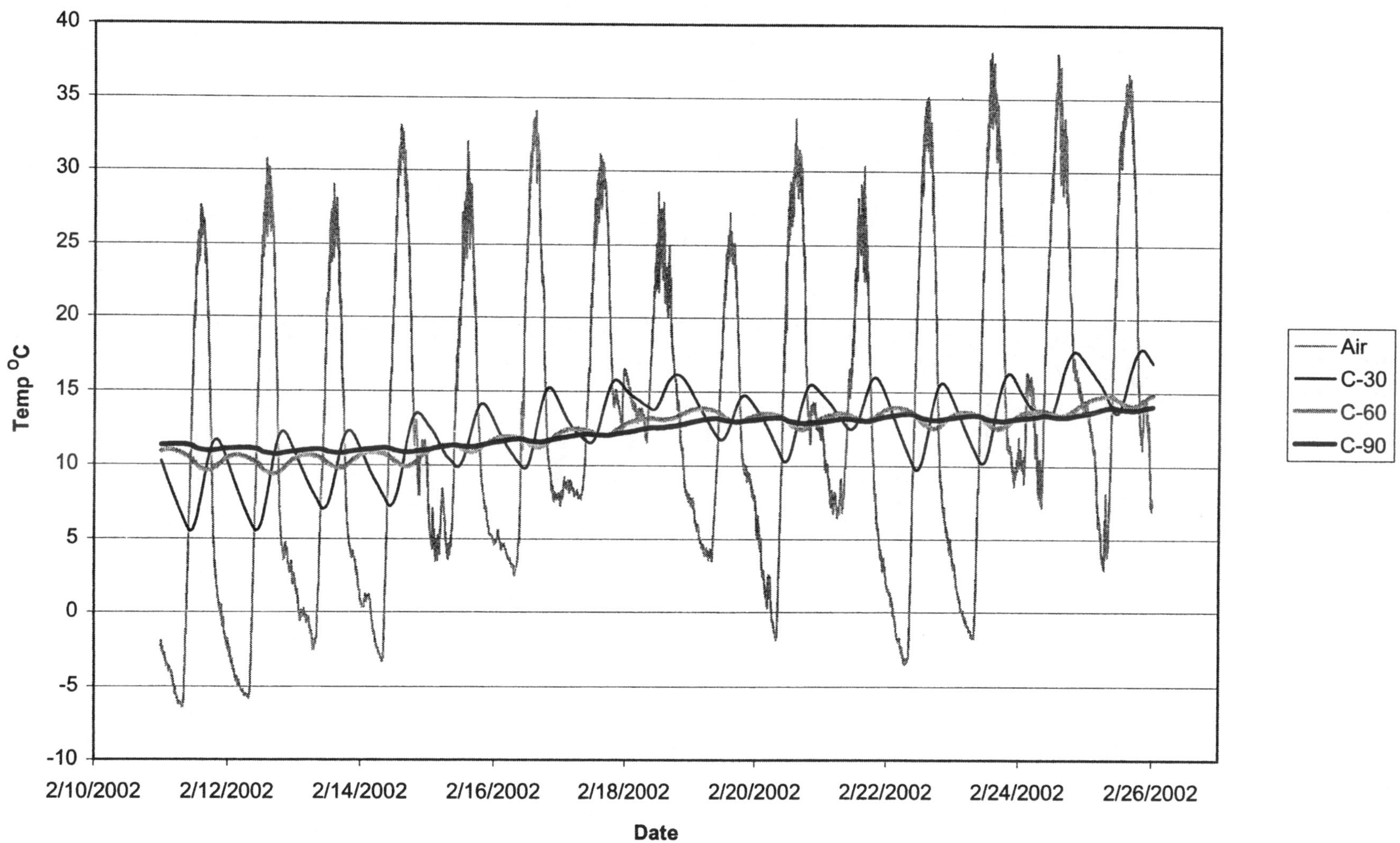

Figure G.4 (continued) Thermocouple data 1/28/02 through 3/22/02

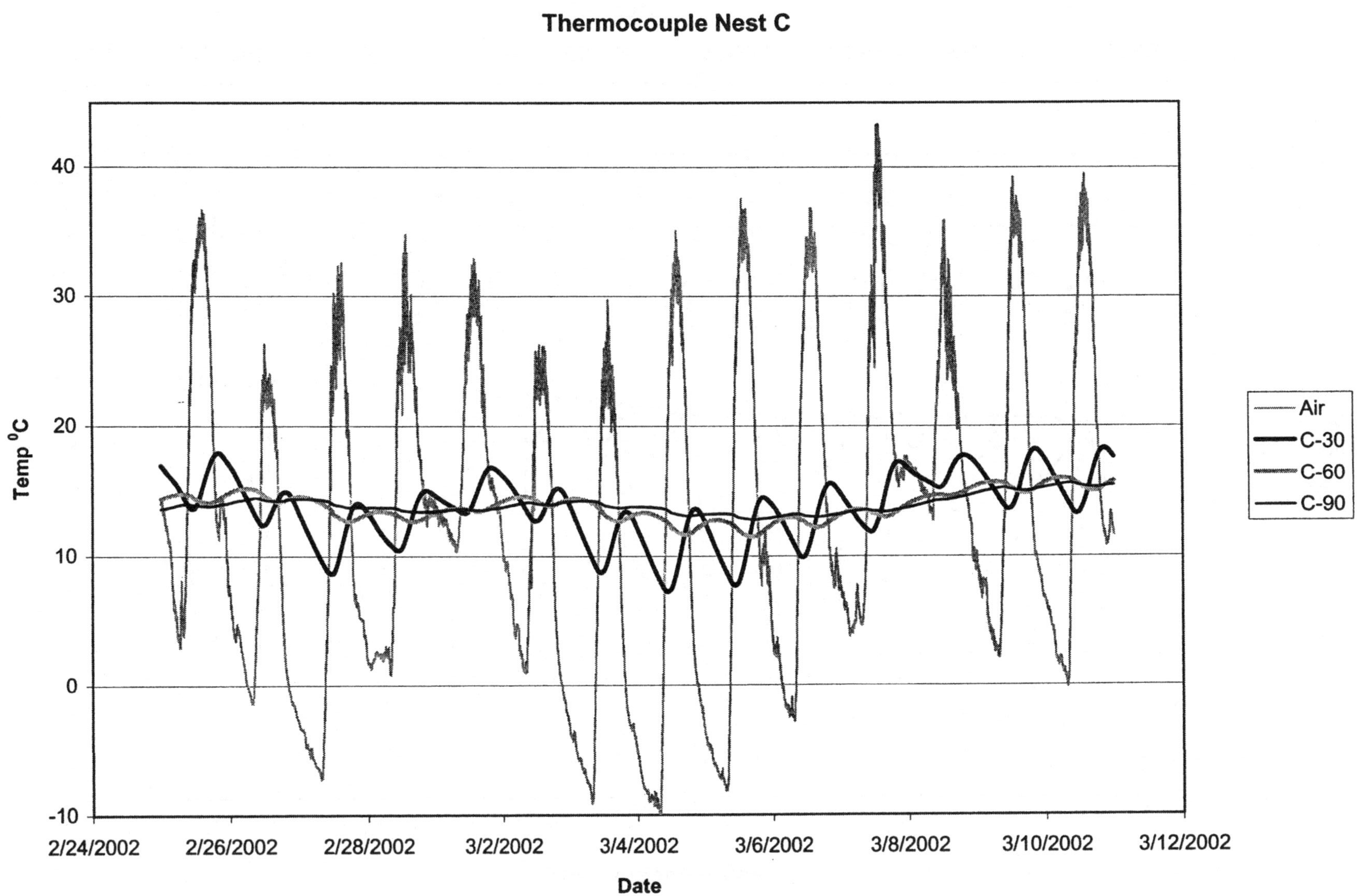

Figure G.4 (continued) Thermocouple data 1/28/02 through 3/22/02

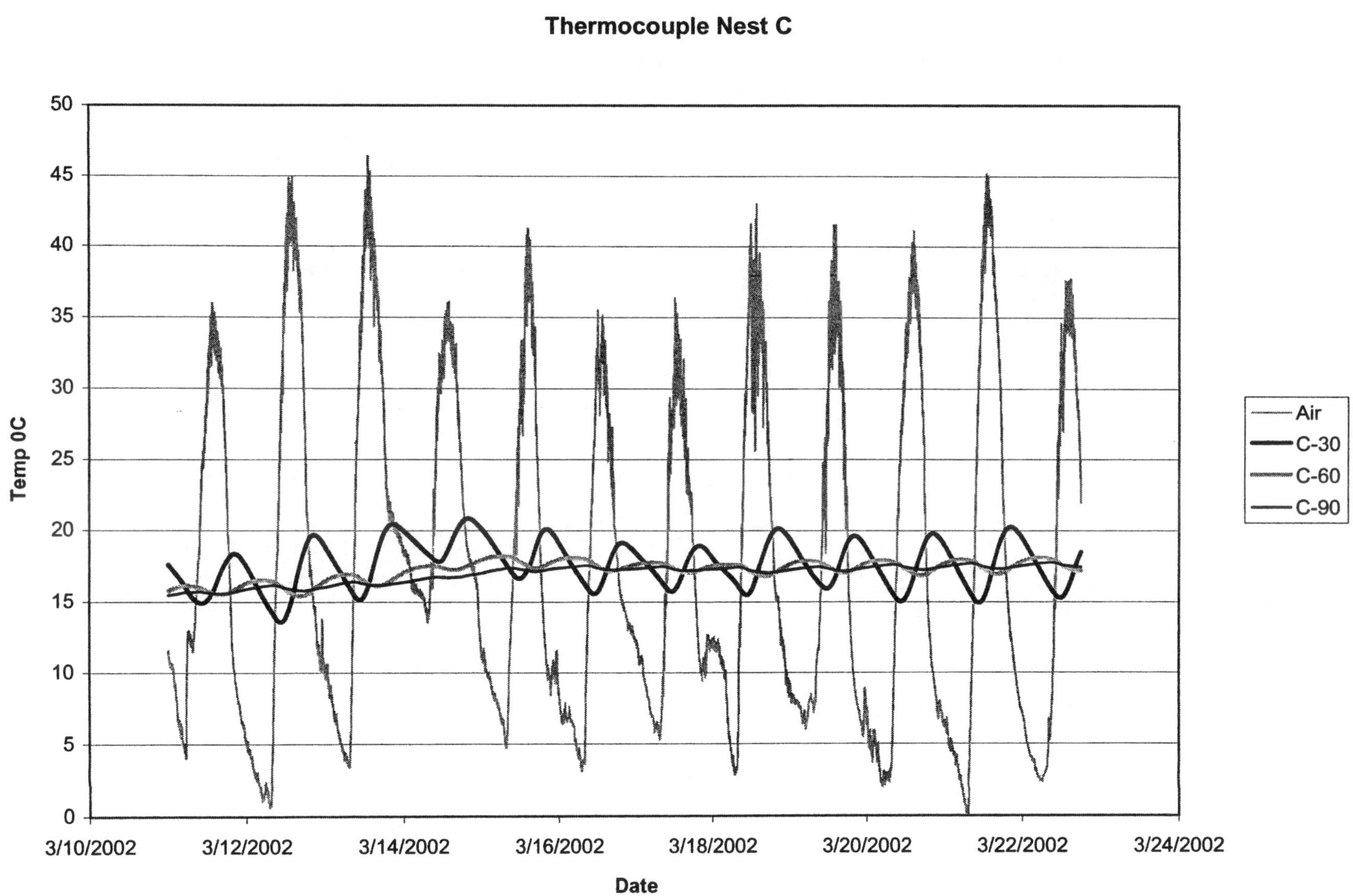

Figure G.4 (continued) Thermocouple data 1/28/02 through 3/22/02

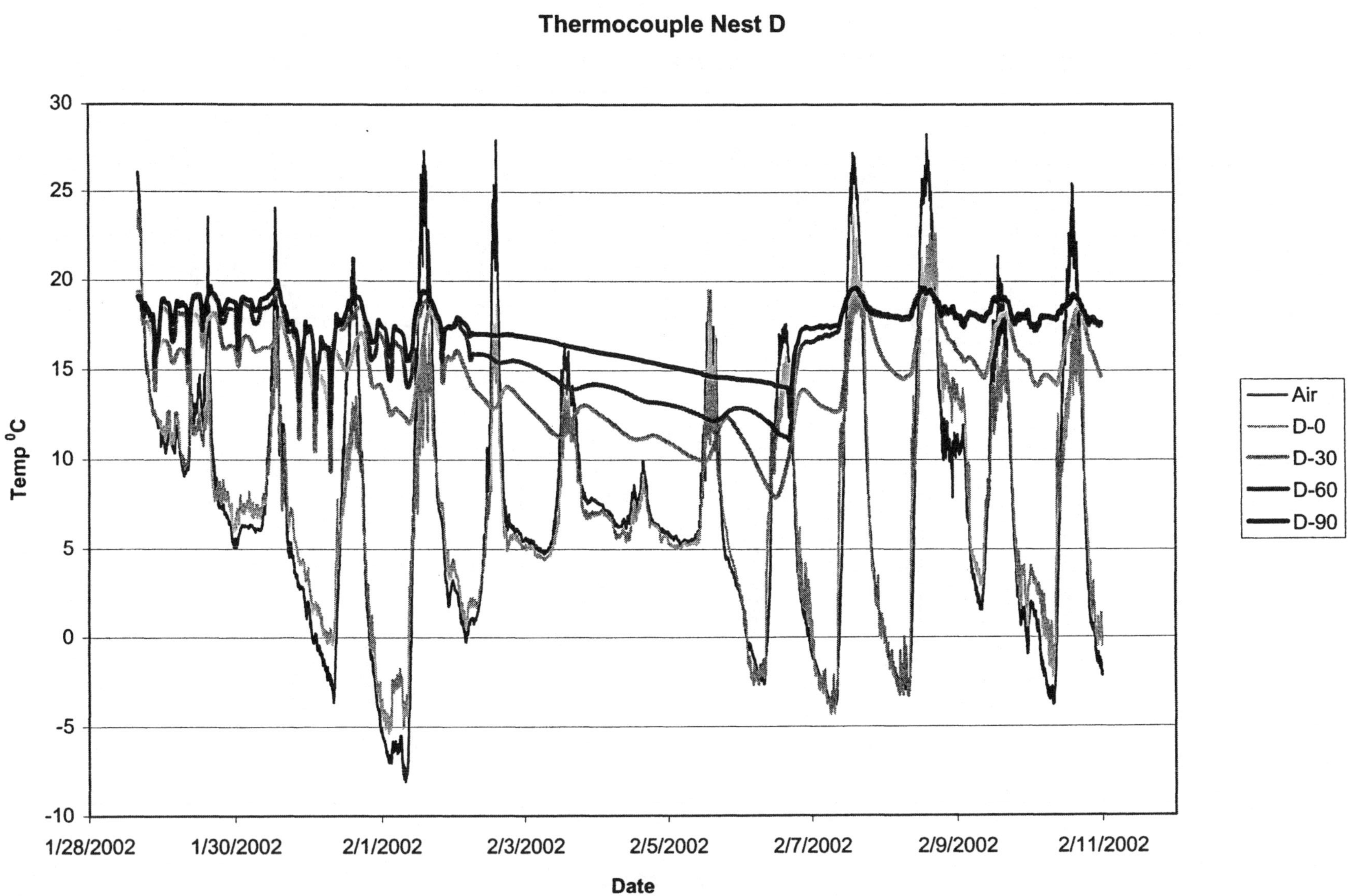

Figure G.4 (continued) Thermocouple data 1/28/02 through 3/22/02

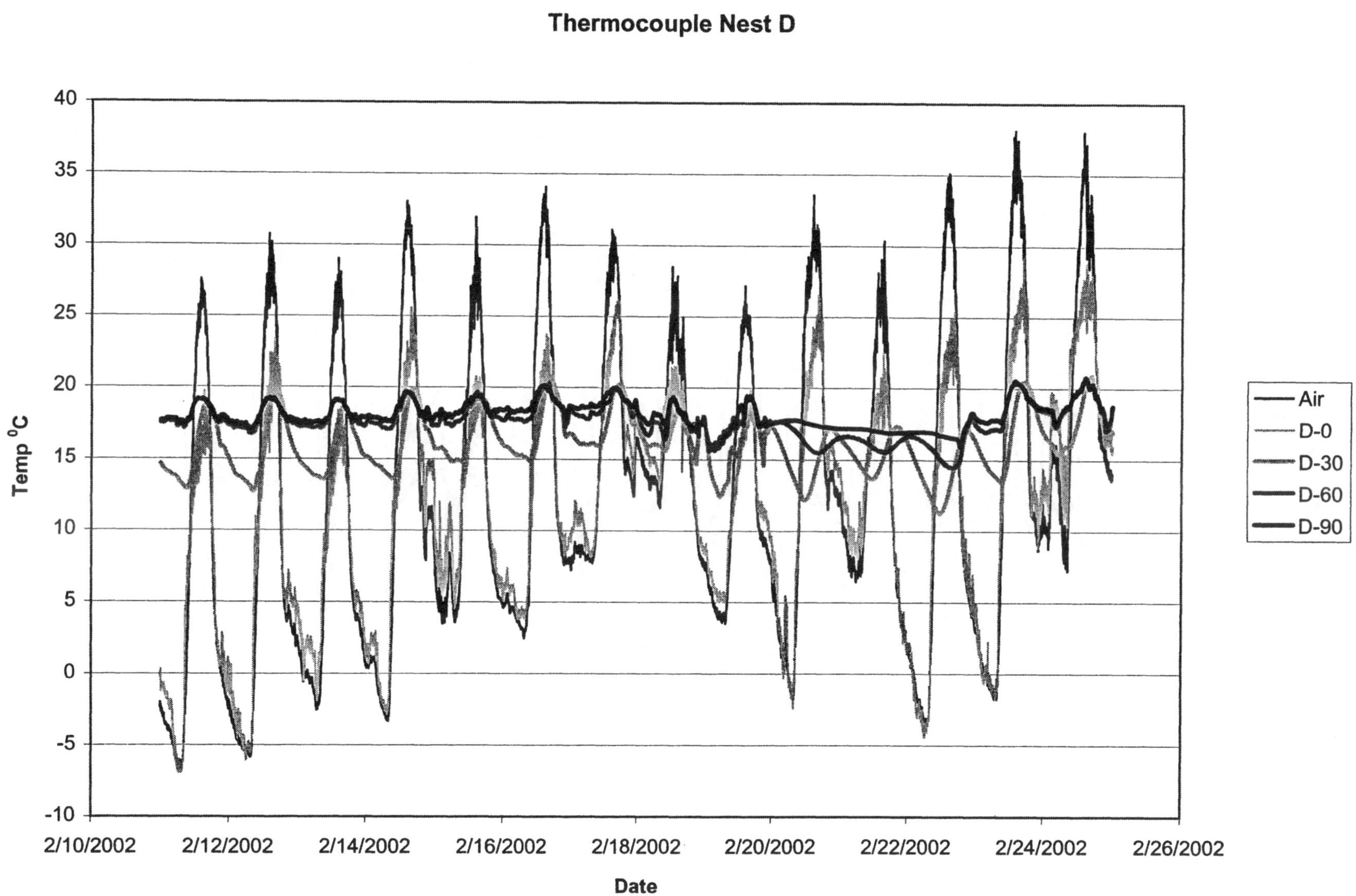

Figure G.4 (continued) Thermocouple data 1/28/02 through 3/22/02

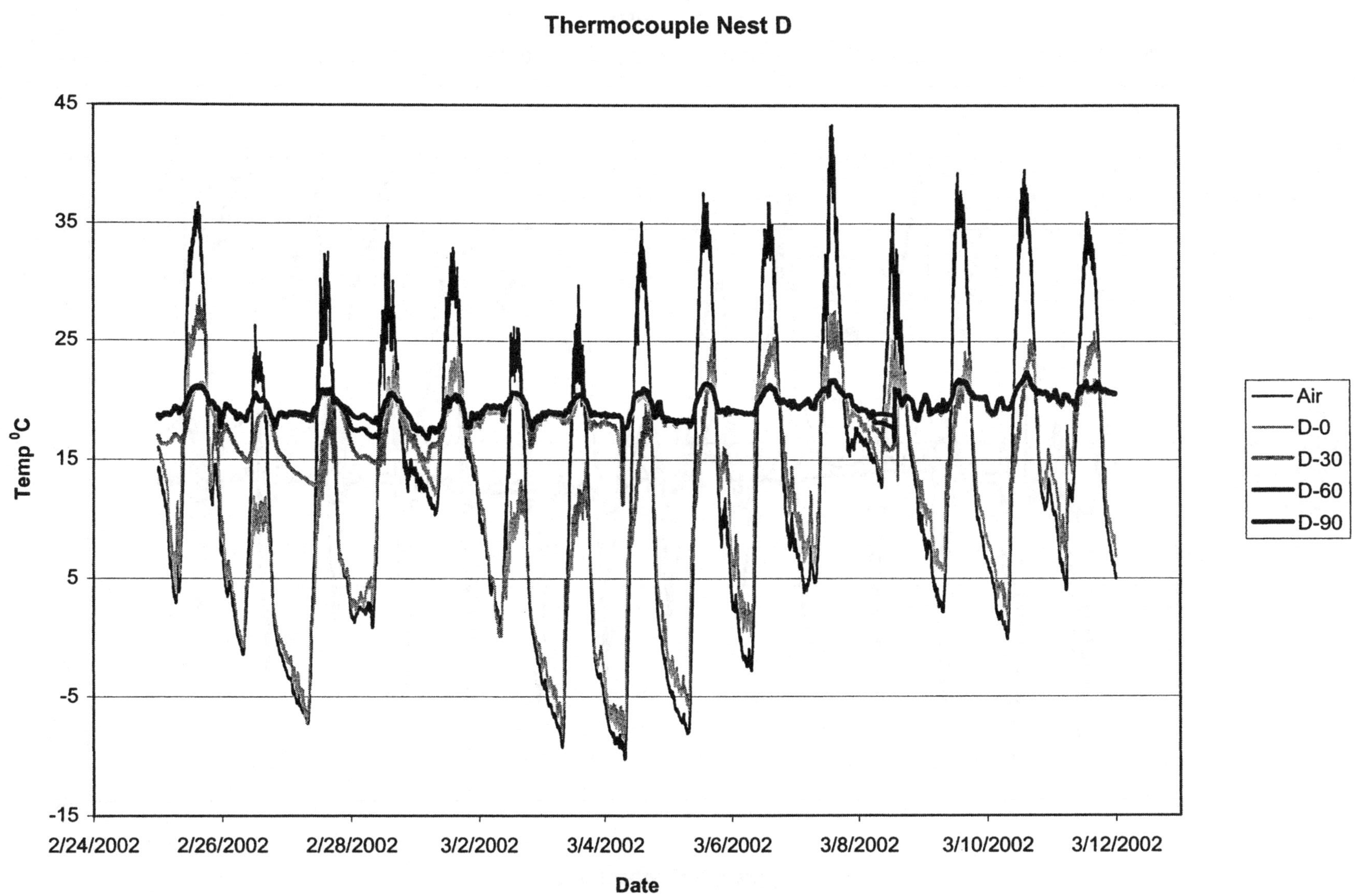

Figure G.4 (continued) Thermocouple data 1/28/02 through 3/22/02

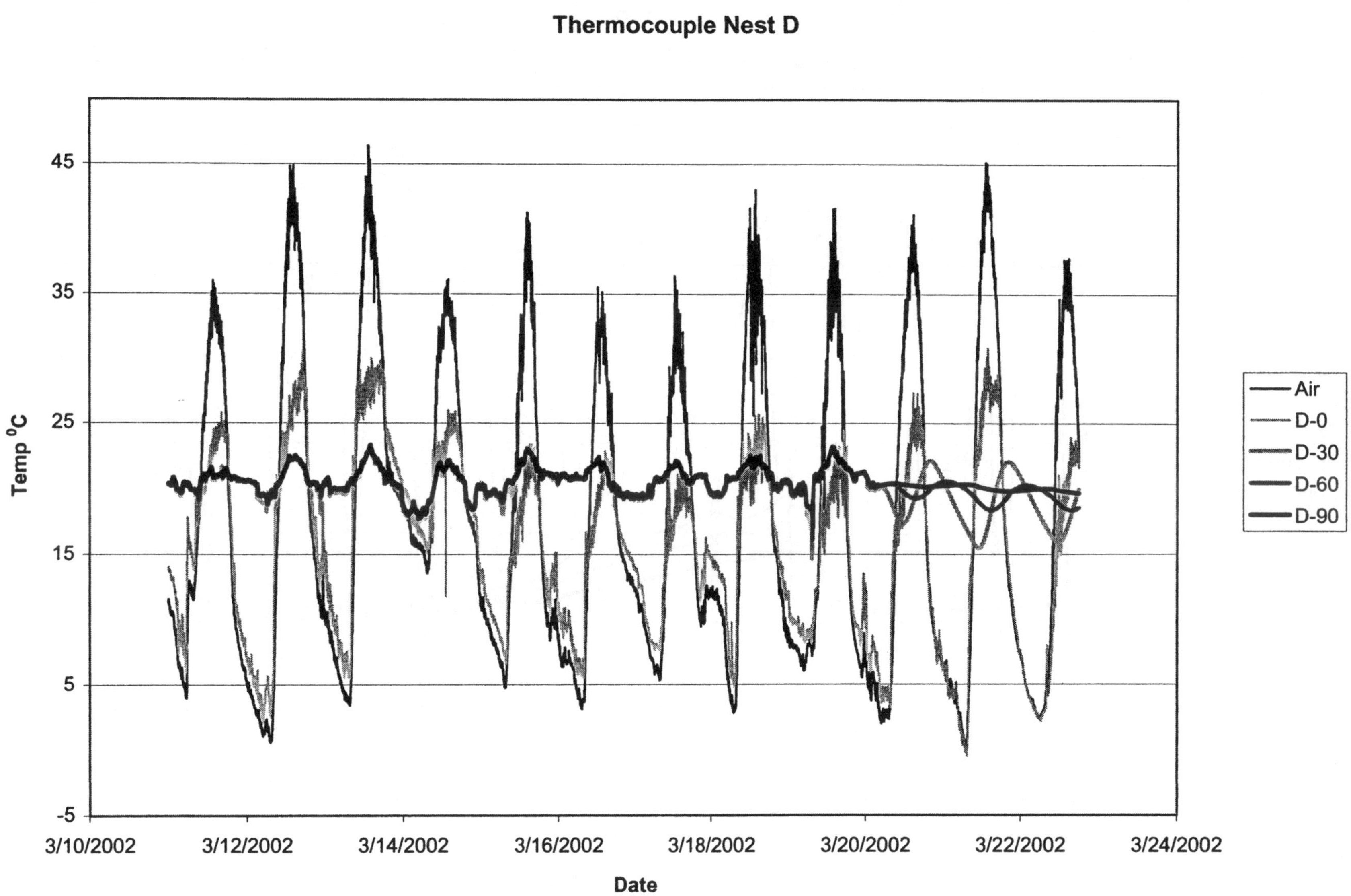

Figure G.4 (continued) Thermocouple data 1/28/02 through 3/22/02

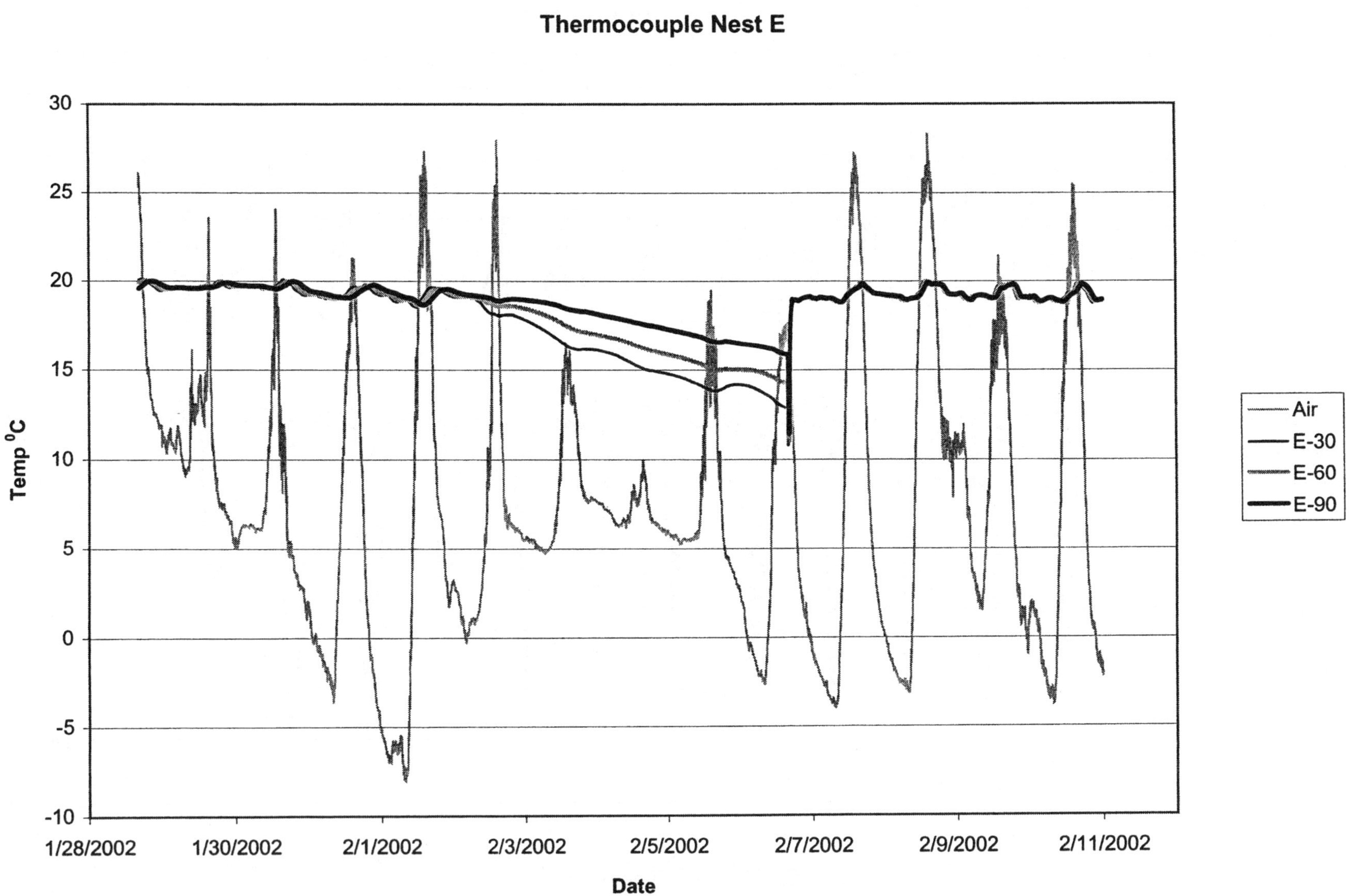

Figure G.4 (continued) Thermocouple data 1/28/02 through 3/22/02

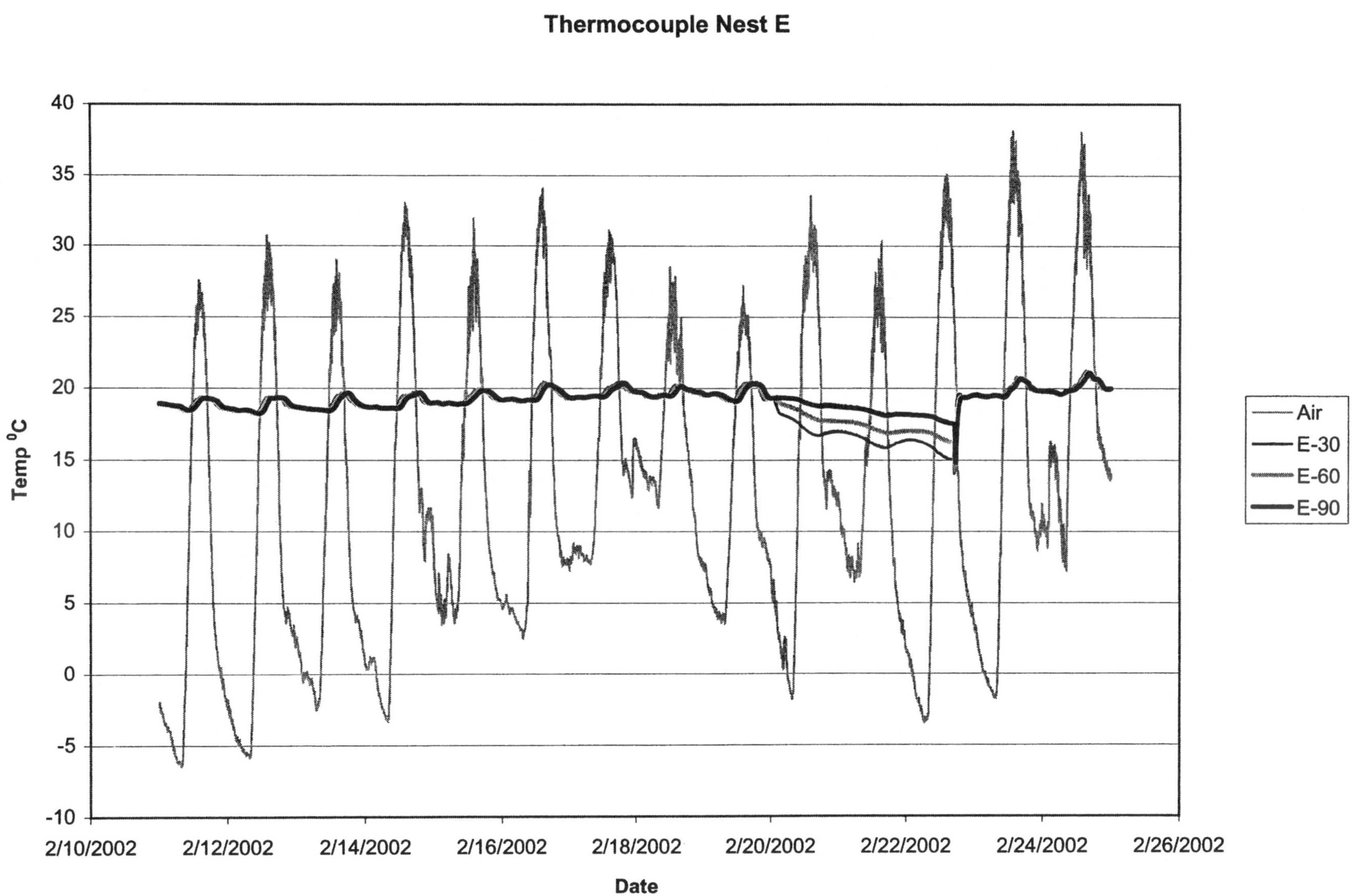

Figure G.4 (continued) Thermocouple data 1/28/02 through 3/22/02

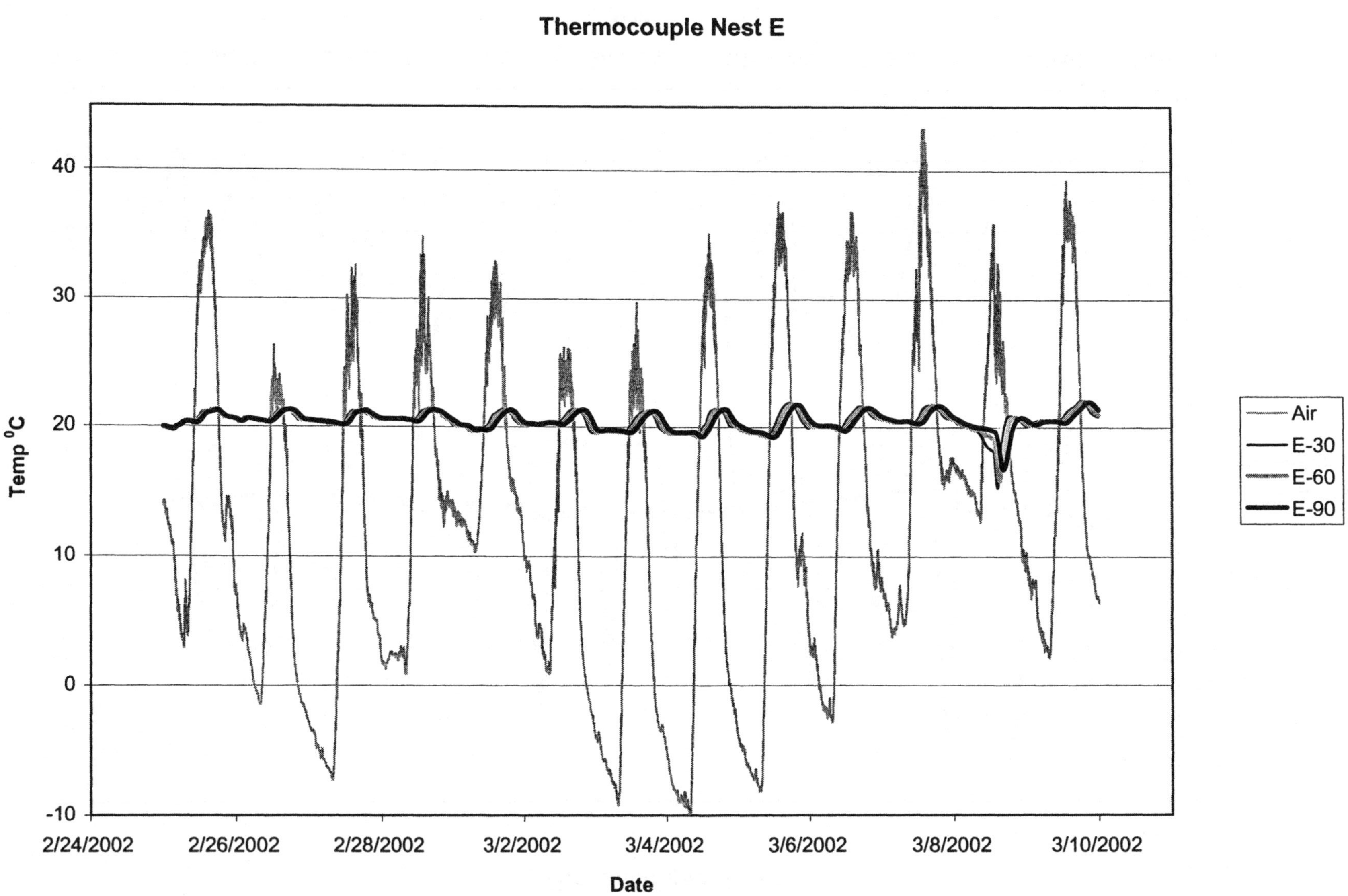

Figure G.4 (continued) Thermocouple data 1/28/02 through 3/22/02

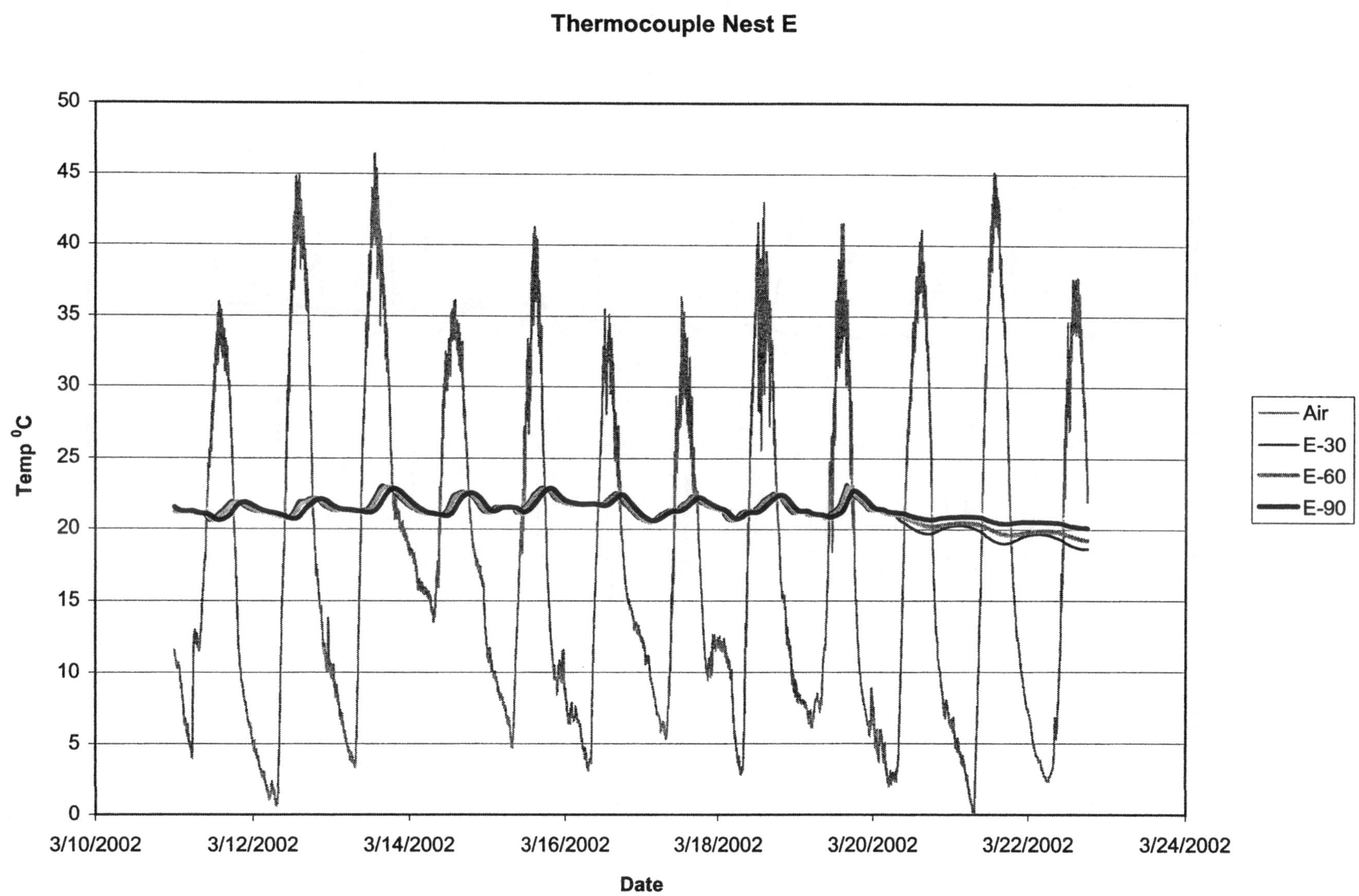

Figure G.4 (continued) Thermocouple data 1/28/02 through 3/22/02

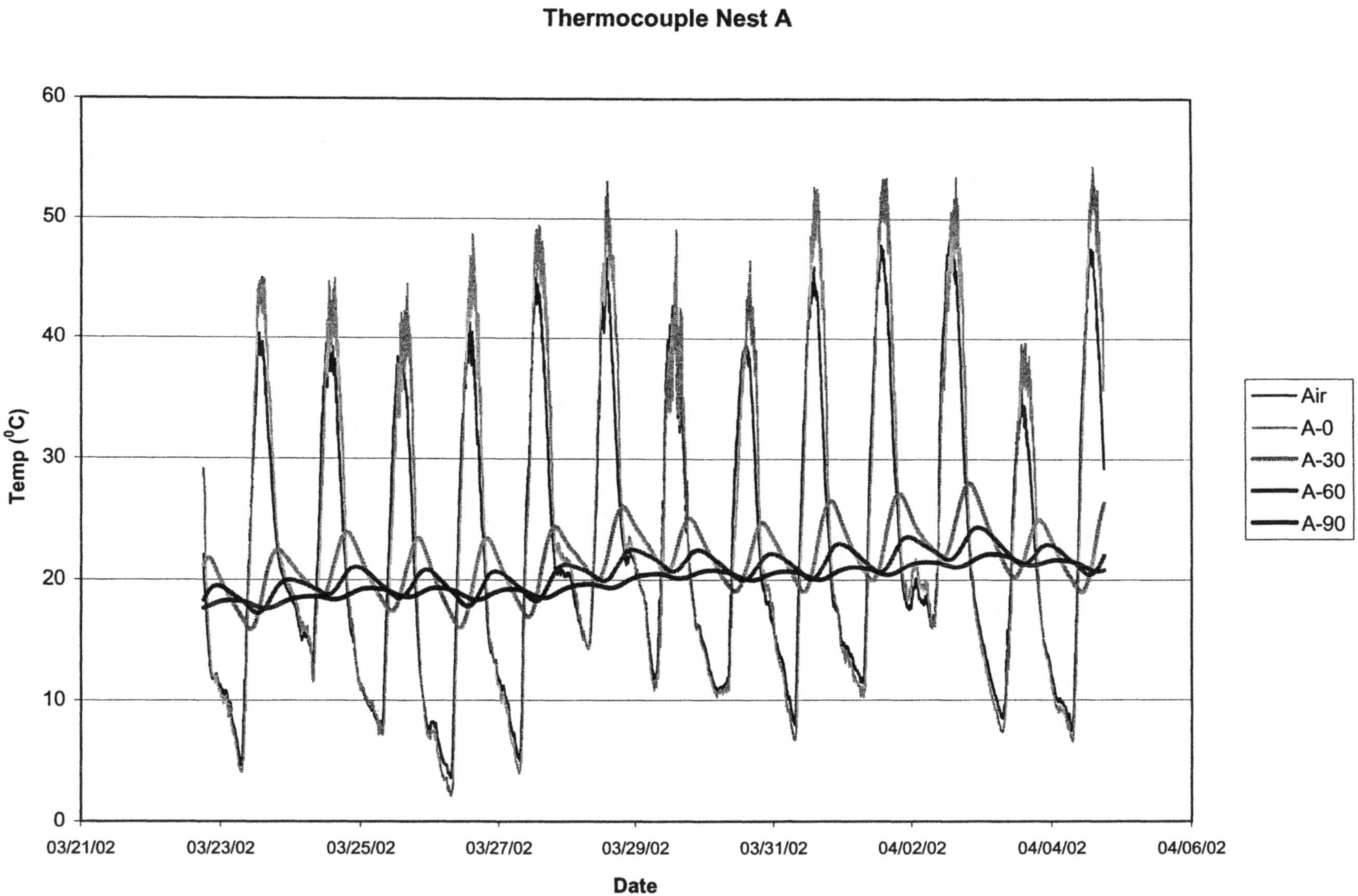

Figure G.5 (continued) Thermocouple data 3/23/02 through 5/15/02

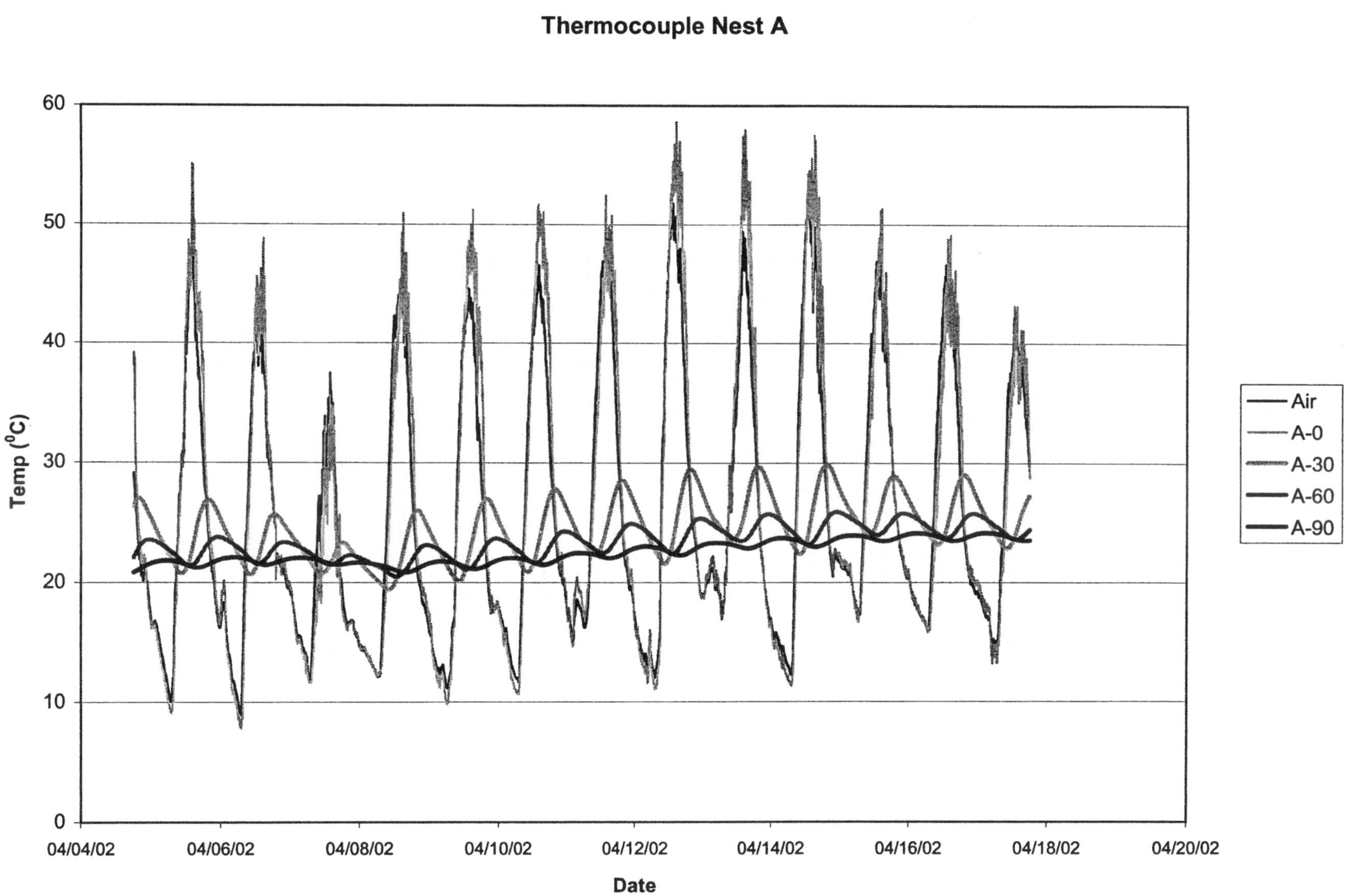

Figure G.5 (continued) Thermocouple data 3/23/02 through 5/15/02

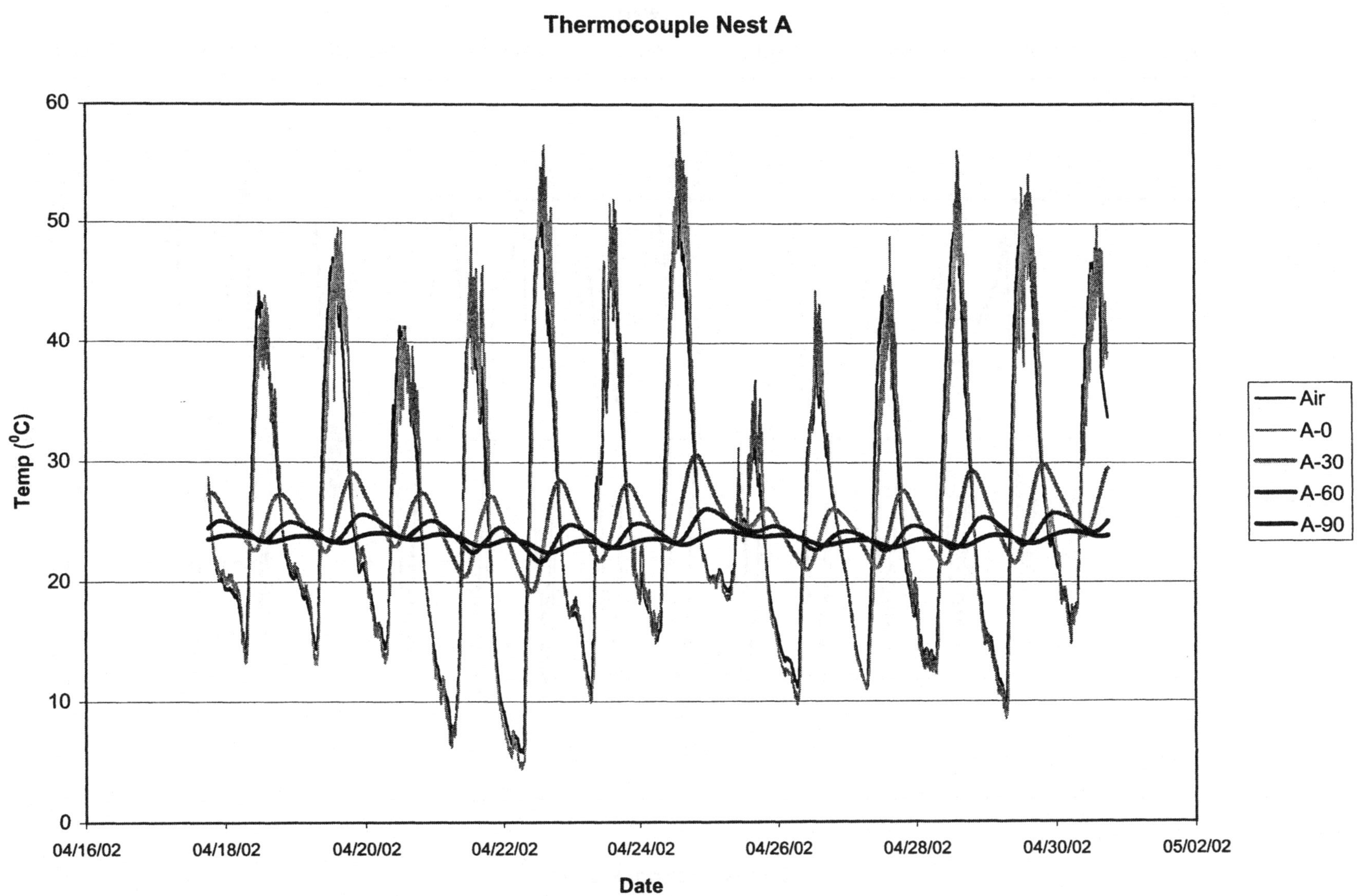

Figure G.5 (continued) Thermocouple data 3/23/02 through 5/15/02

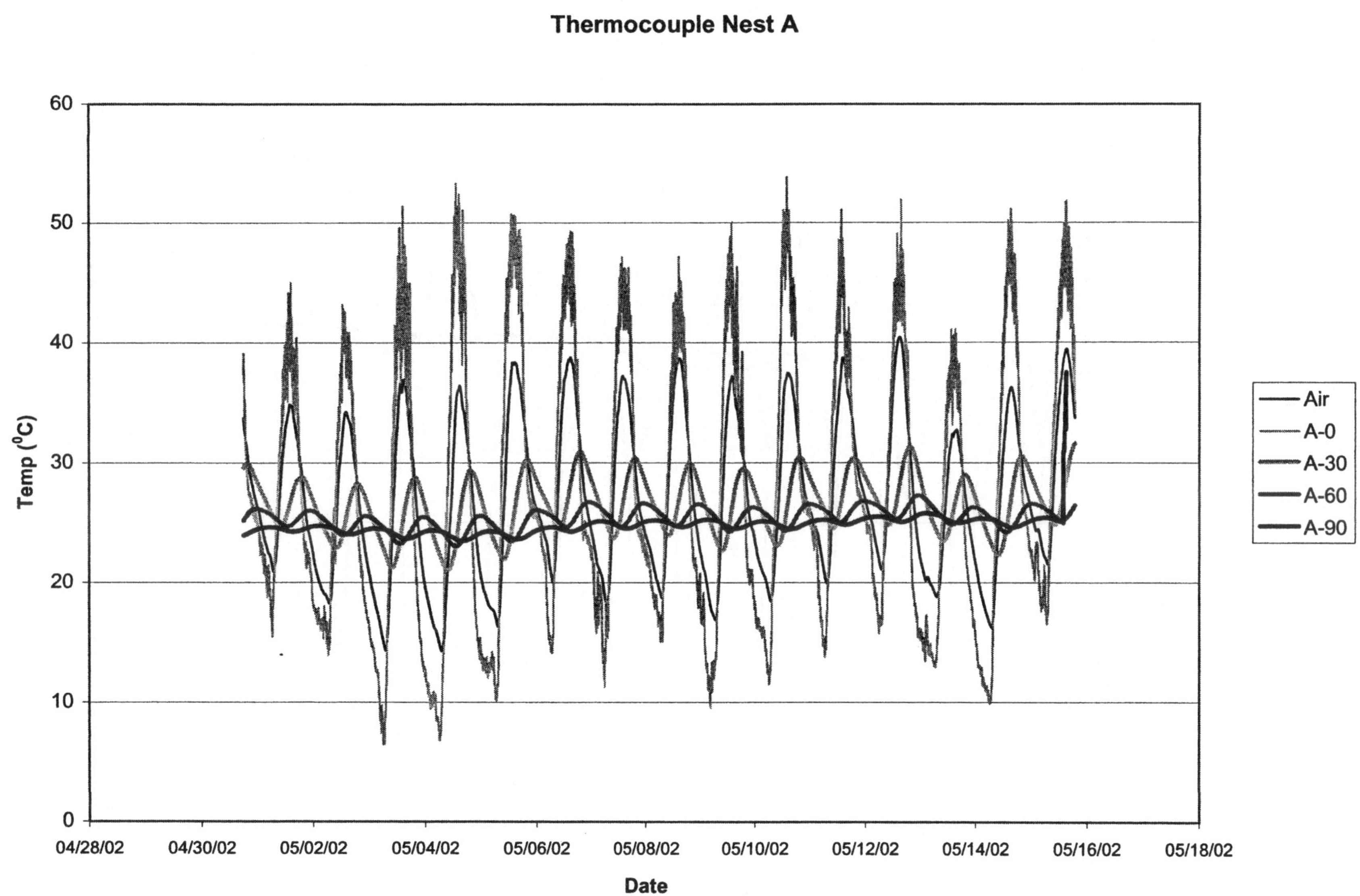

Figure G.5 (continued) Thermocouple data 3/23/02 through 5/15/02

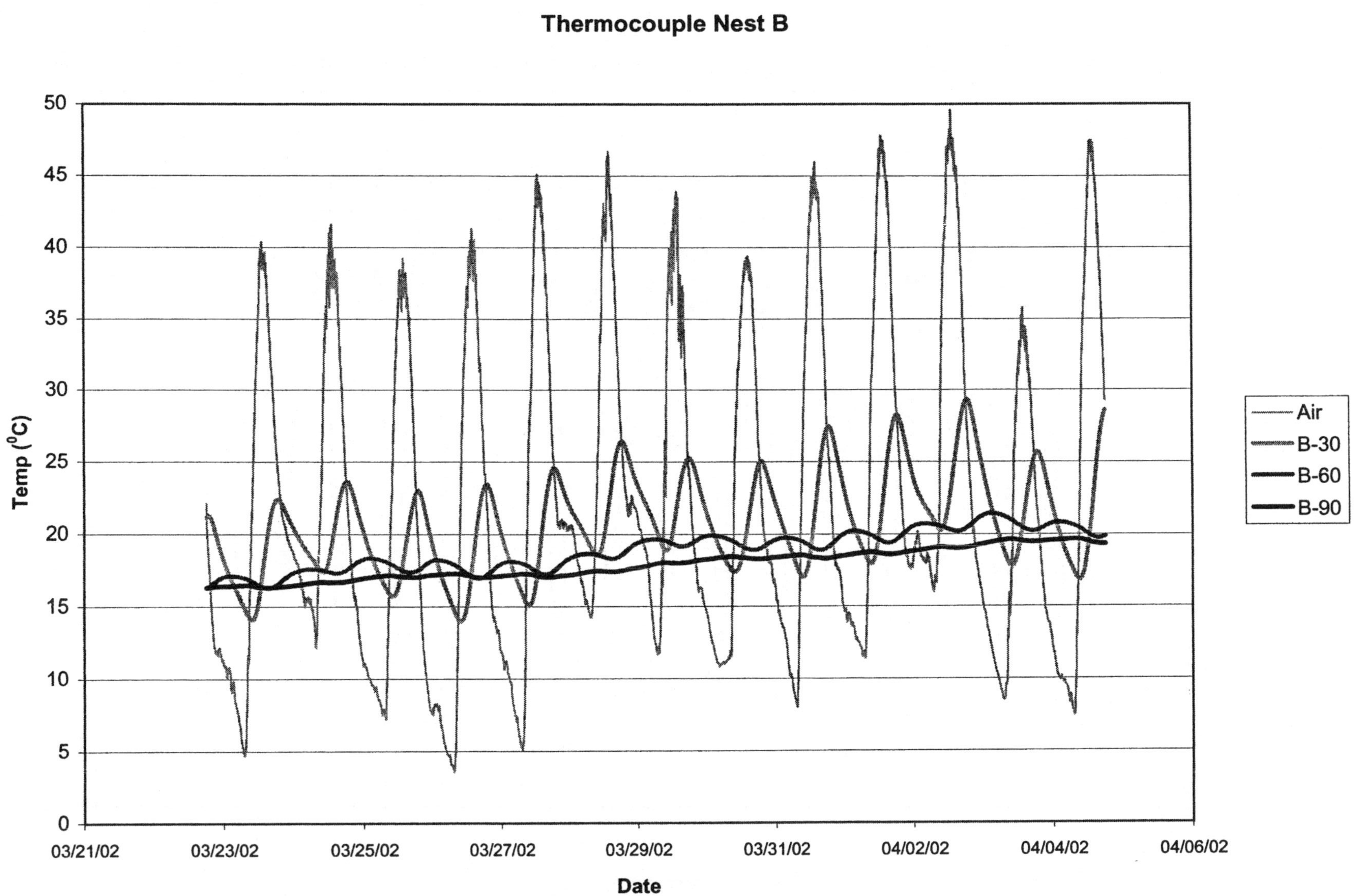

Figure G.5 (continued) Thermocouple data 3/23/02 through 5/15/02

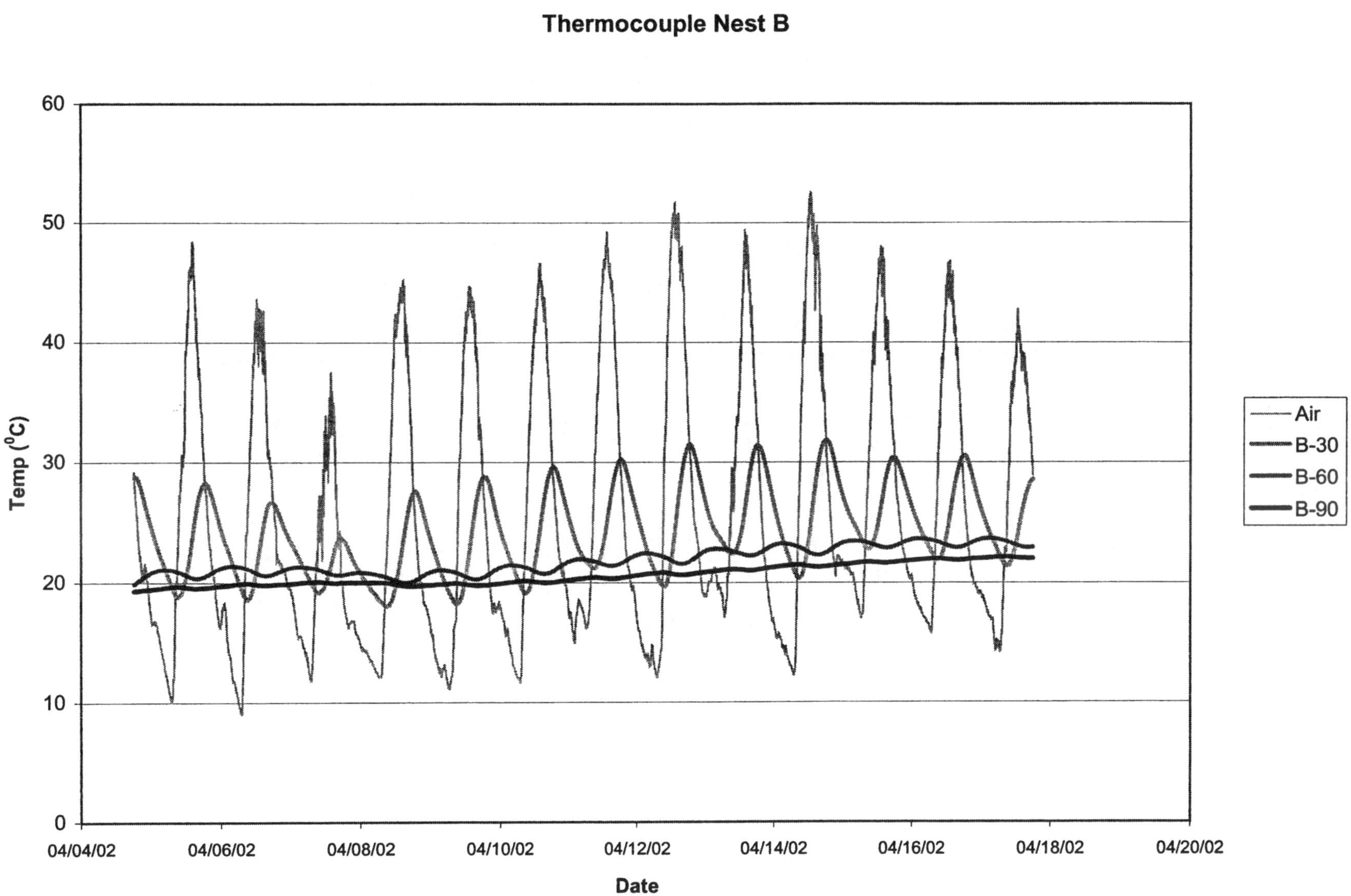

Figure G.5 (continued) Thermocouple data 3/23/02 through 5/15/02

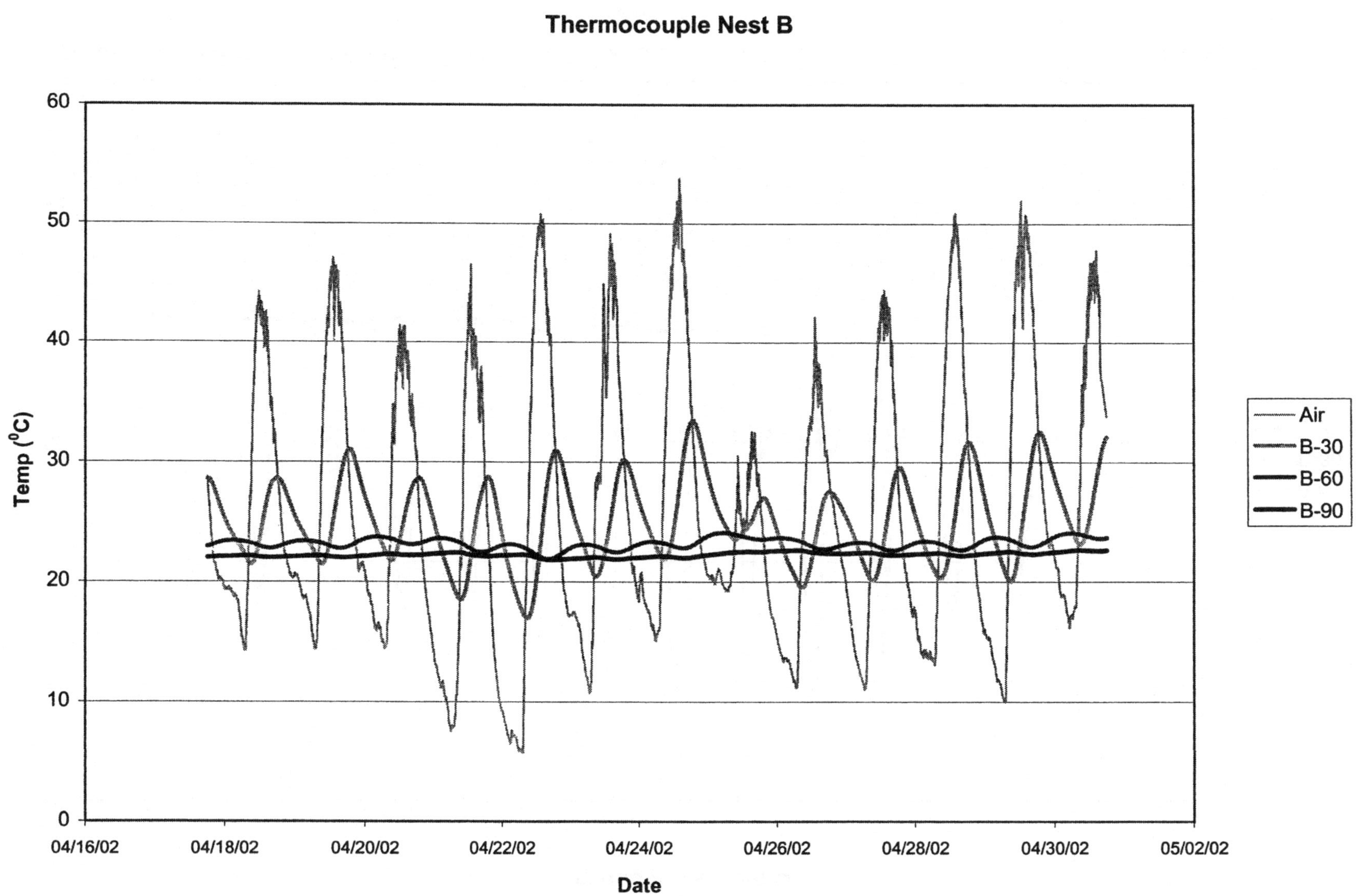

Figure G.5 (continued) Thermocouple data 3/23/02 through 5/15/02

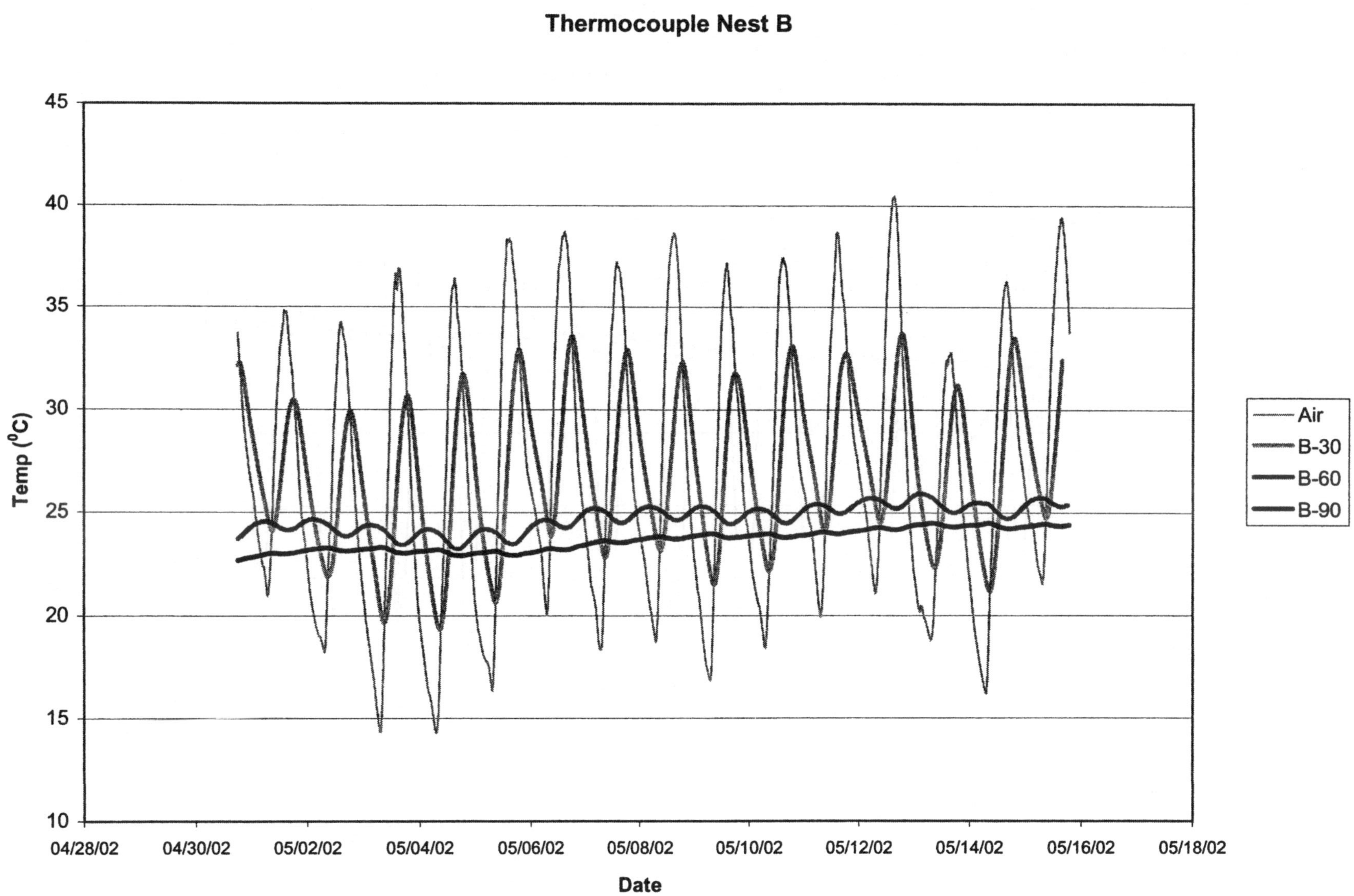

Figure G.5 (continued) Thermocouple data 3/23/02 through 5/15/02

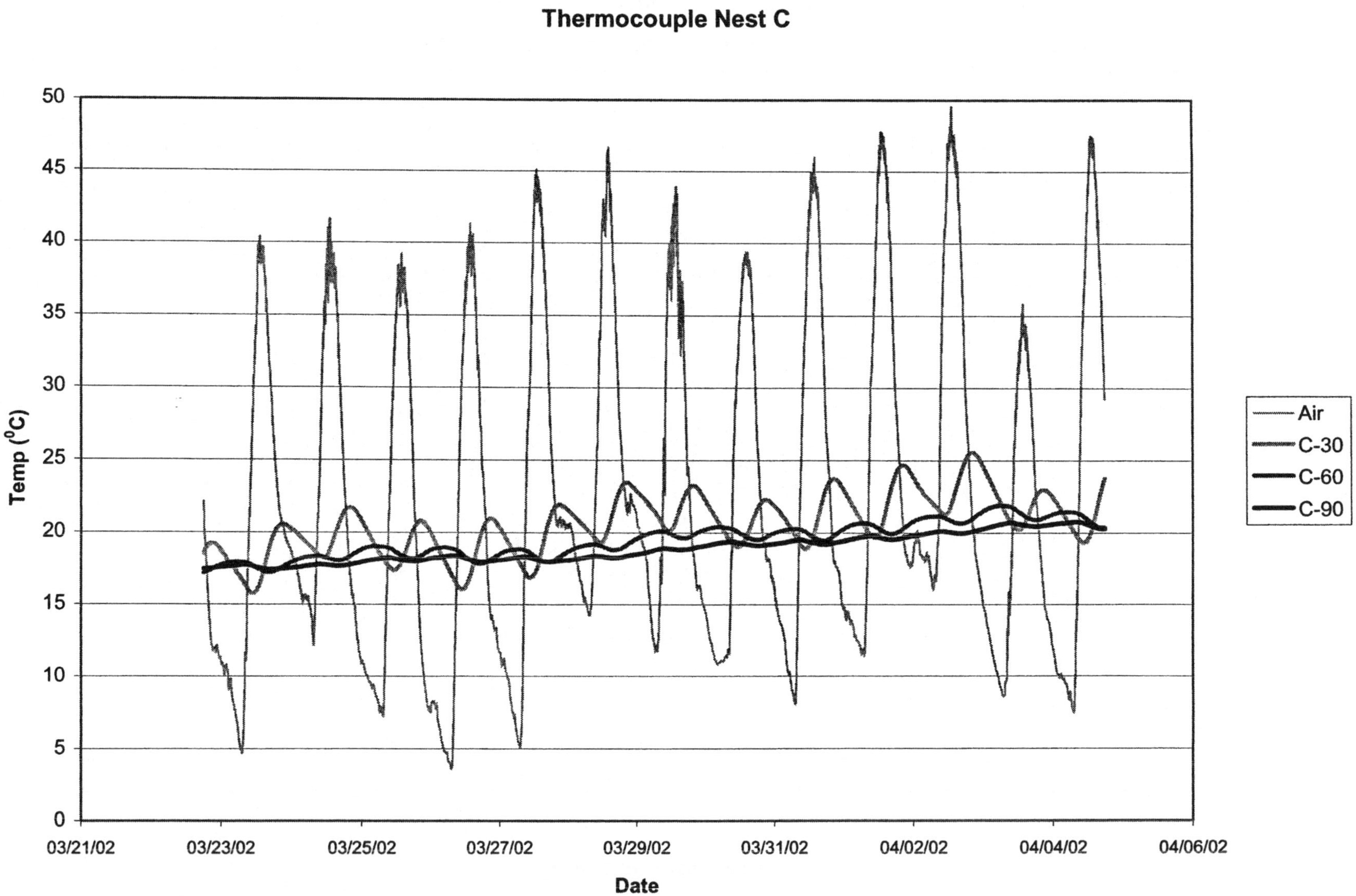

Figure G.5 (continued) Thermocouple data 3/23/02 through 5/15/02

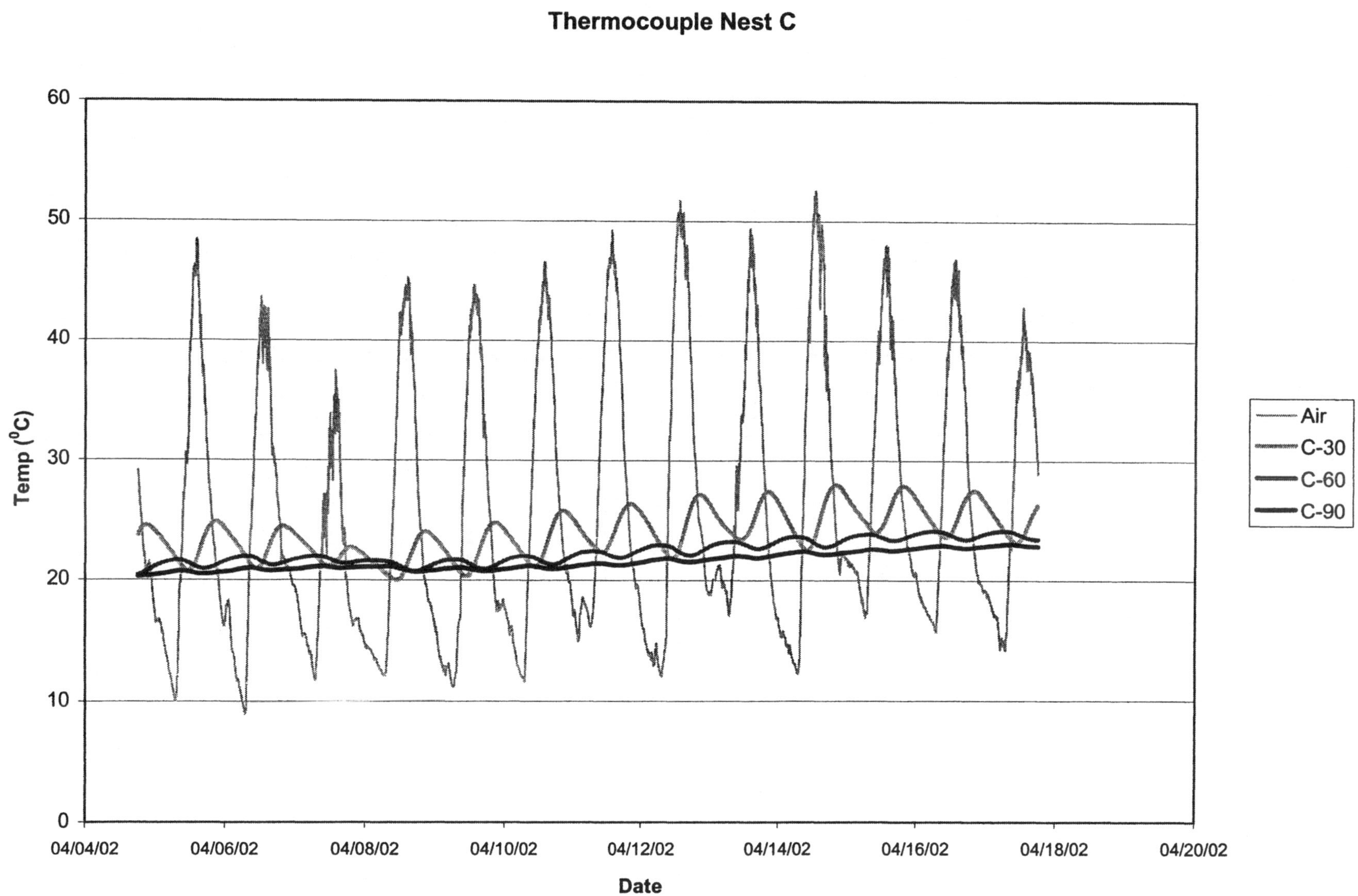

Figure G.5 (continued) Thermocouple data 3/23/02 through 5/15/02

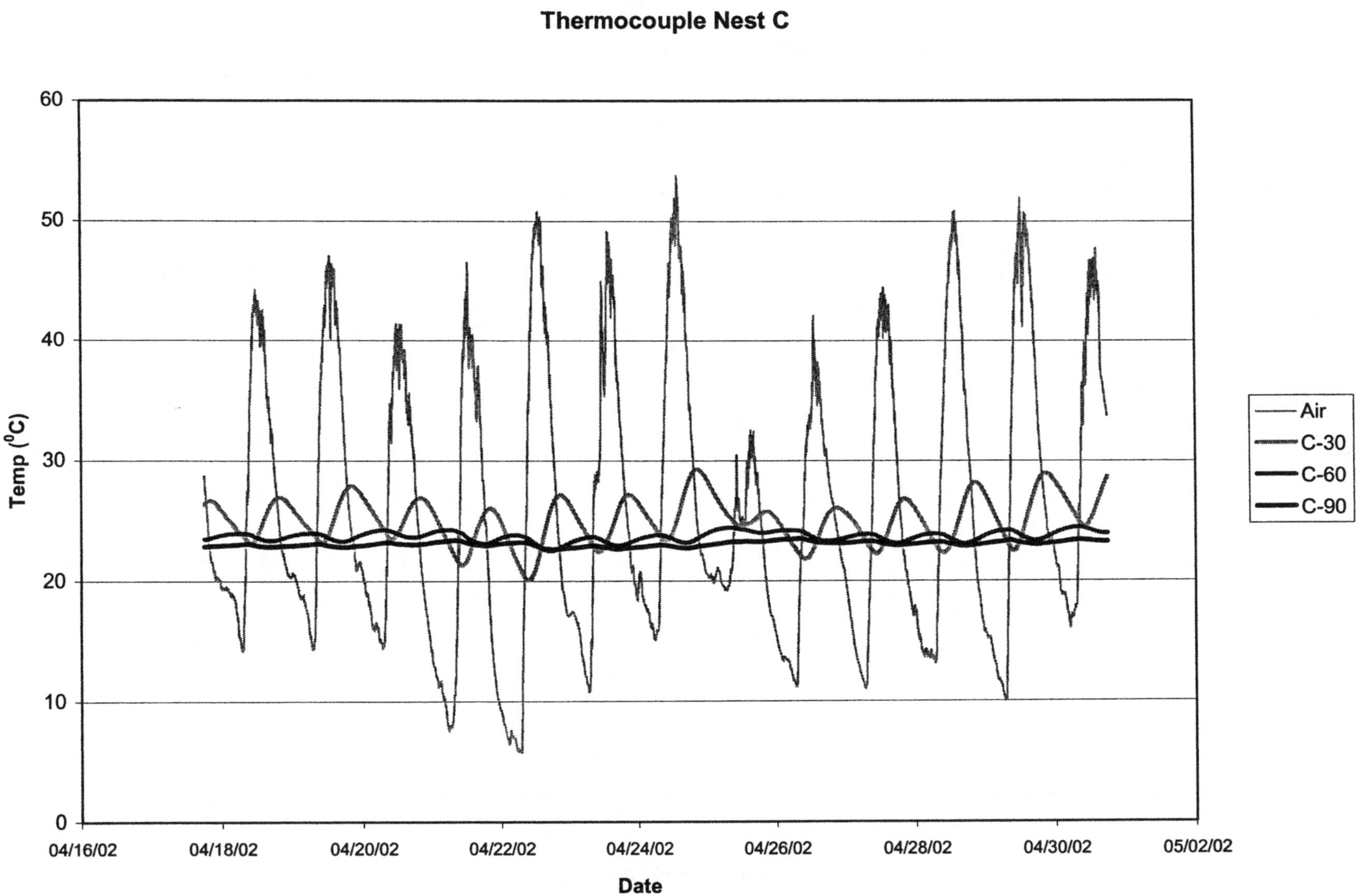

Figure G.5 (continued) Thermocouple data 3/23/02 through 5/15/02

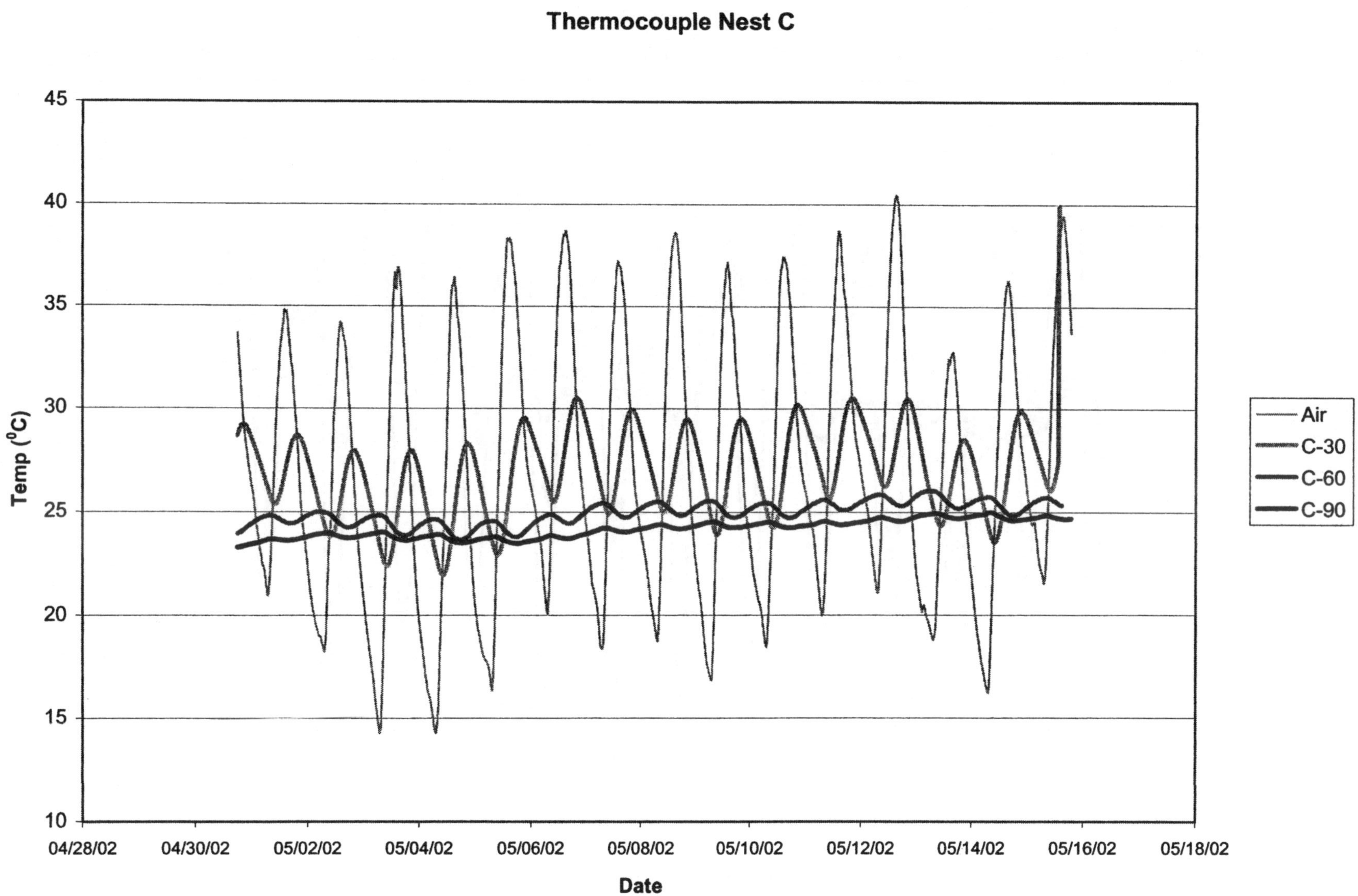

Figure G.5 (continued) Thermocouple data 3/23/02 through 5/15/02

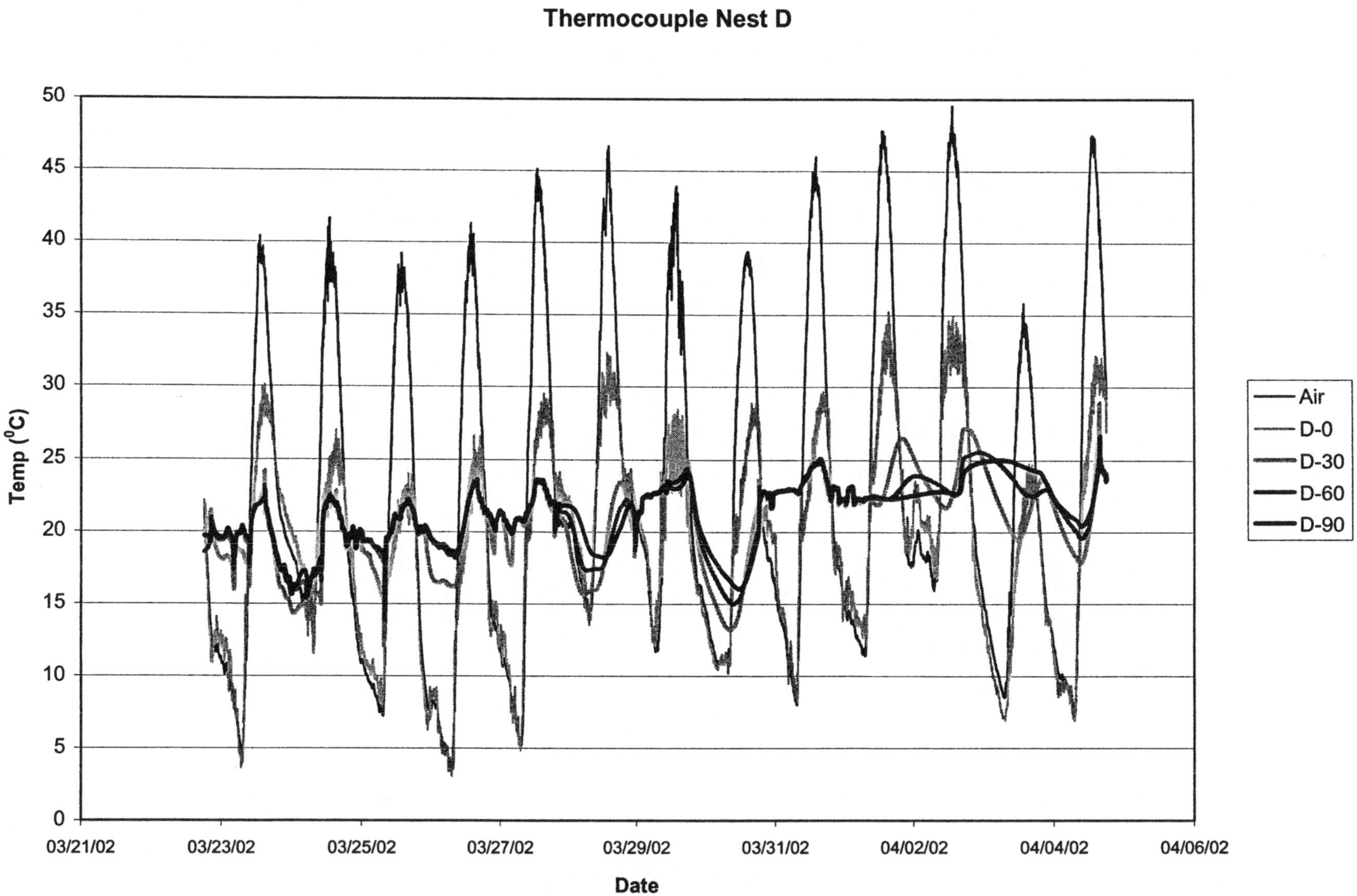

Figure G.5 (continued) Thermocouple data 3/23/02 through 5/15/02

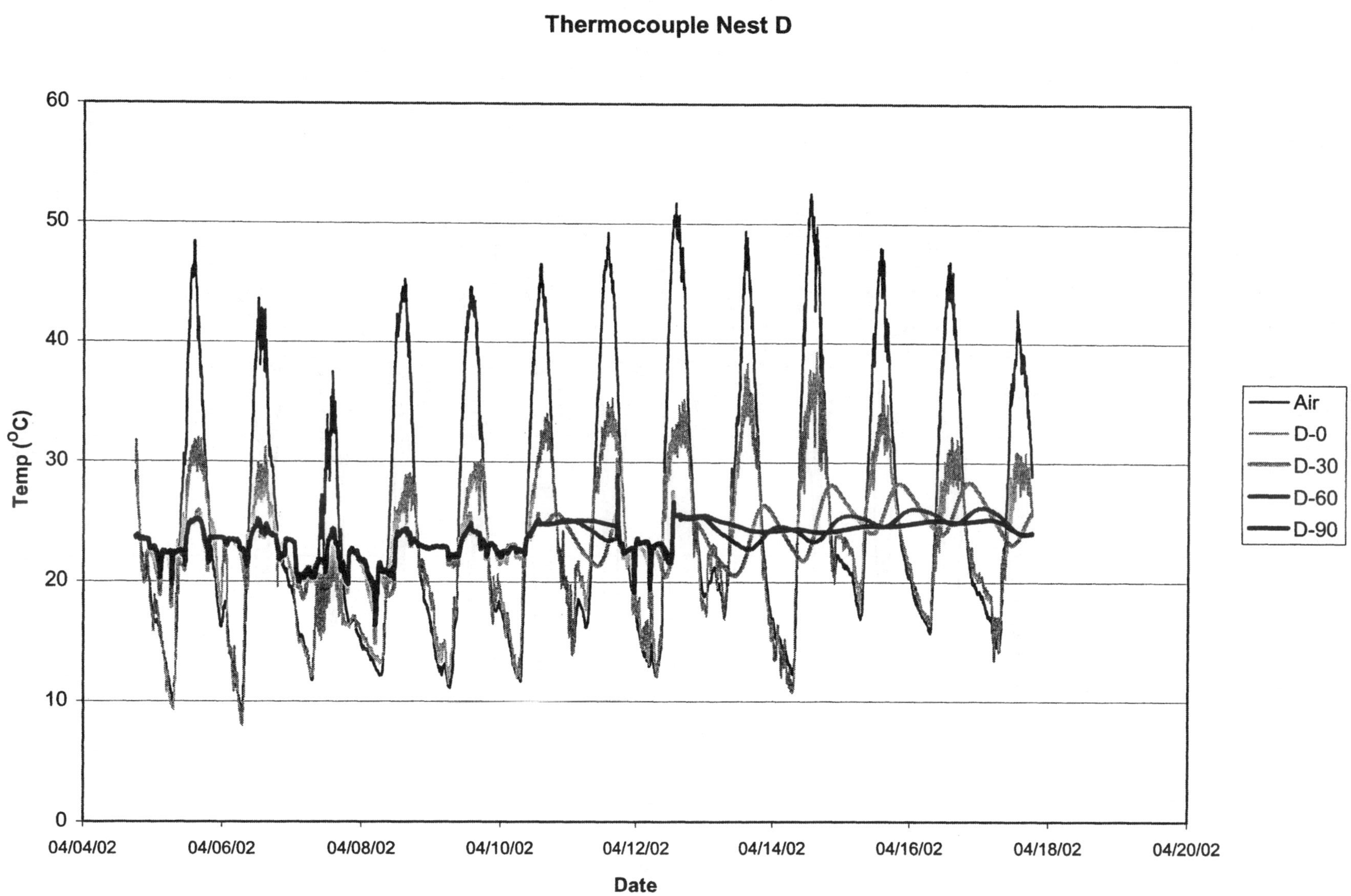

Figure G.5 (continued) Thermocouple data 3/23/02 through 5/15/02

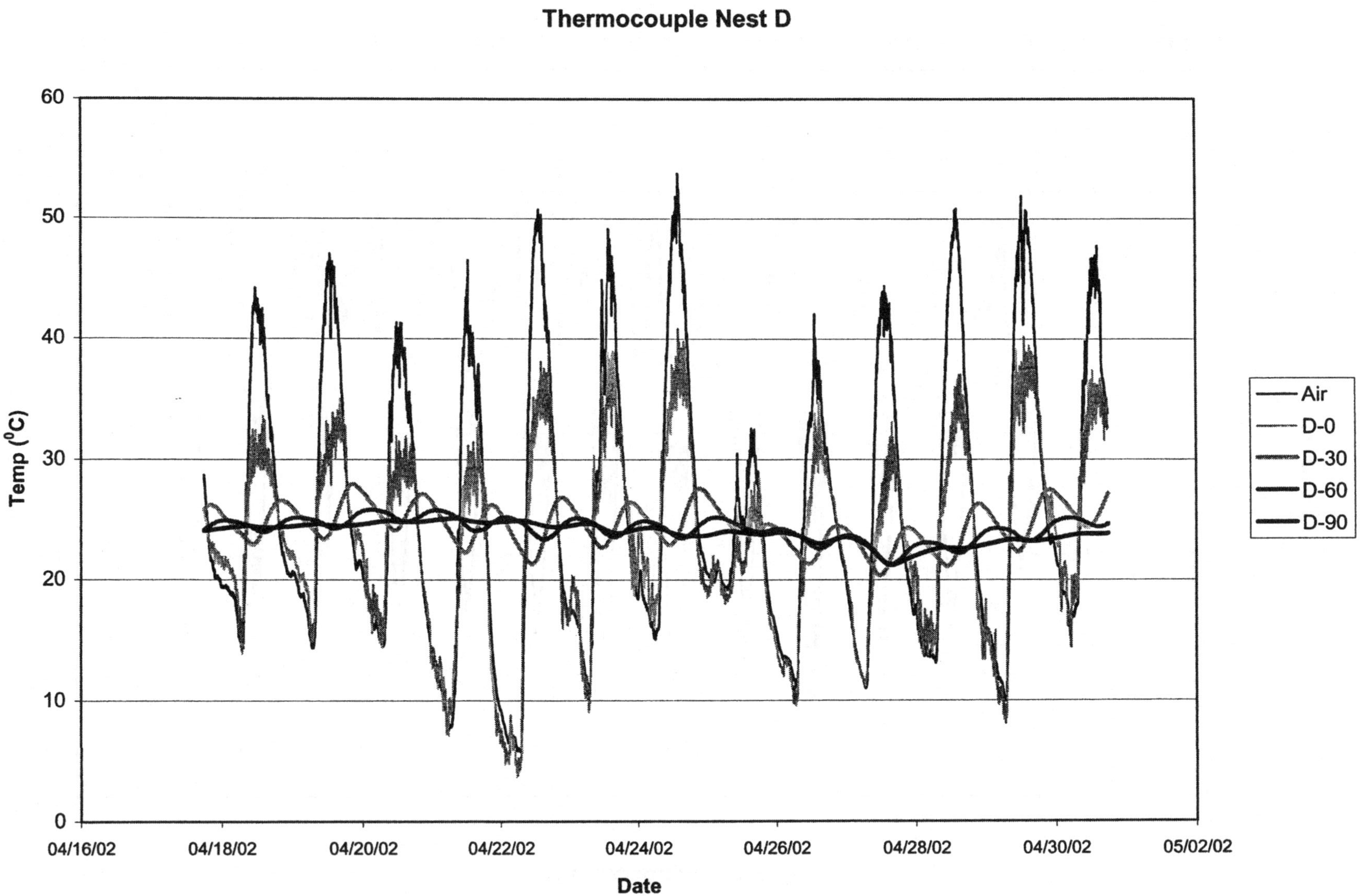

Figure G.5 (continued) Thermocouple data 3/23/02 through 5/15/02

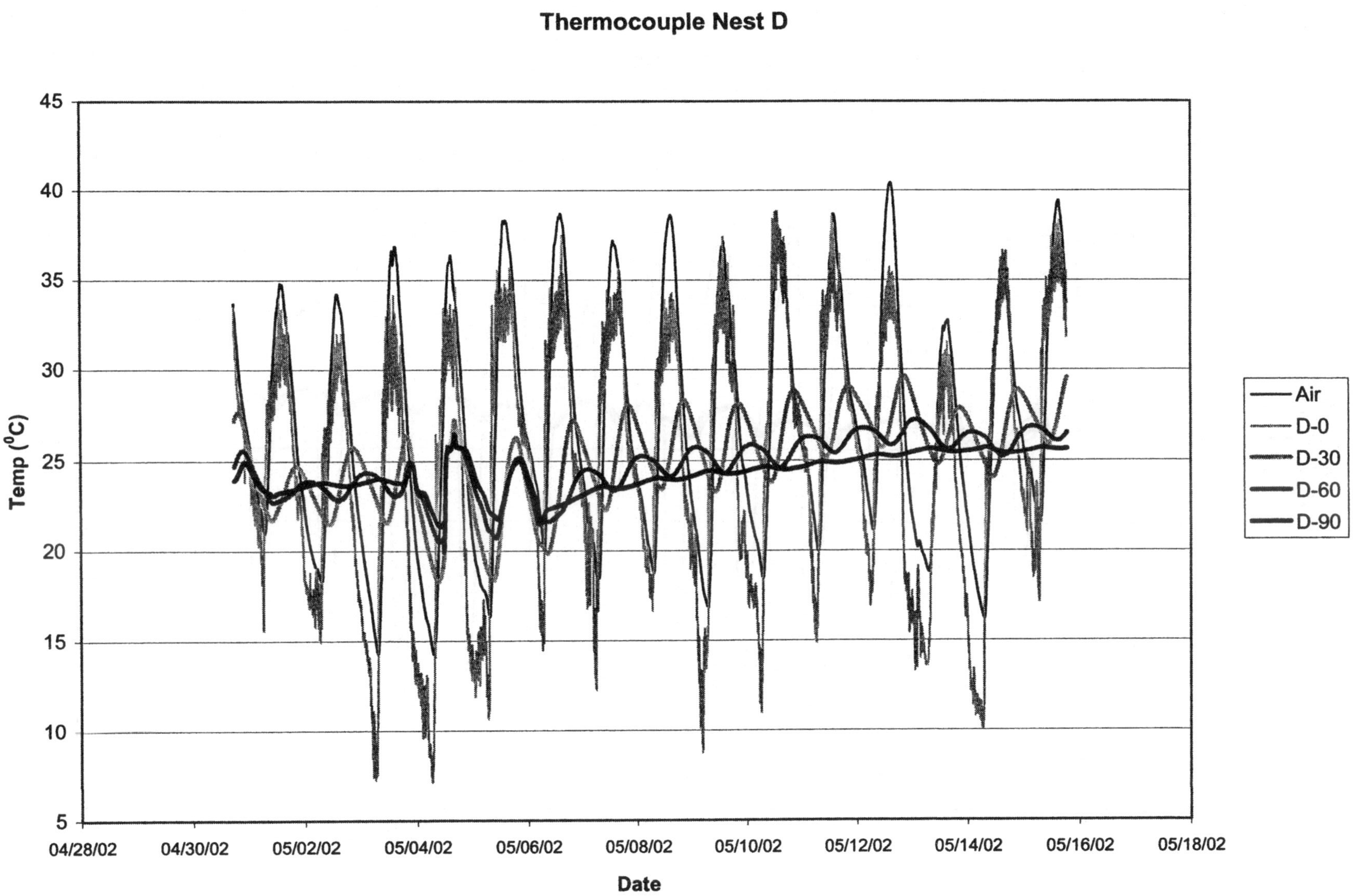

Figure G.5 (continued) Thermocouple data 3/23/02 through 5/15/02

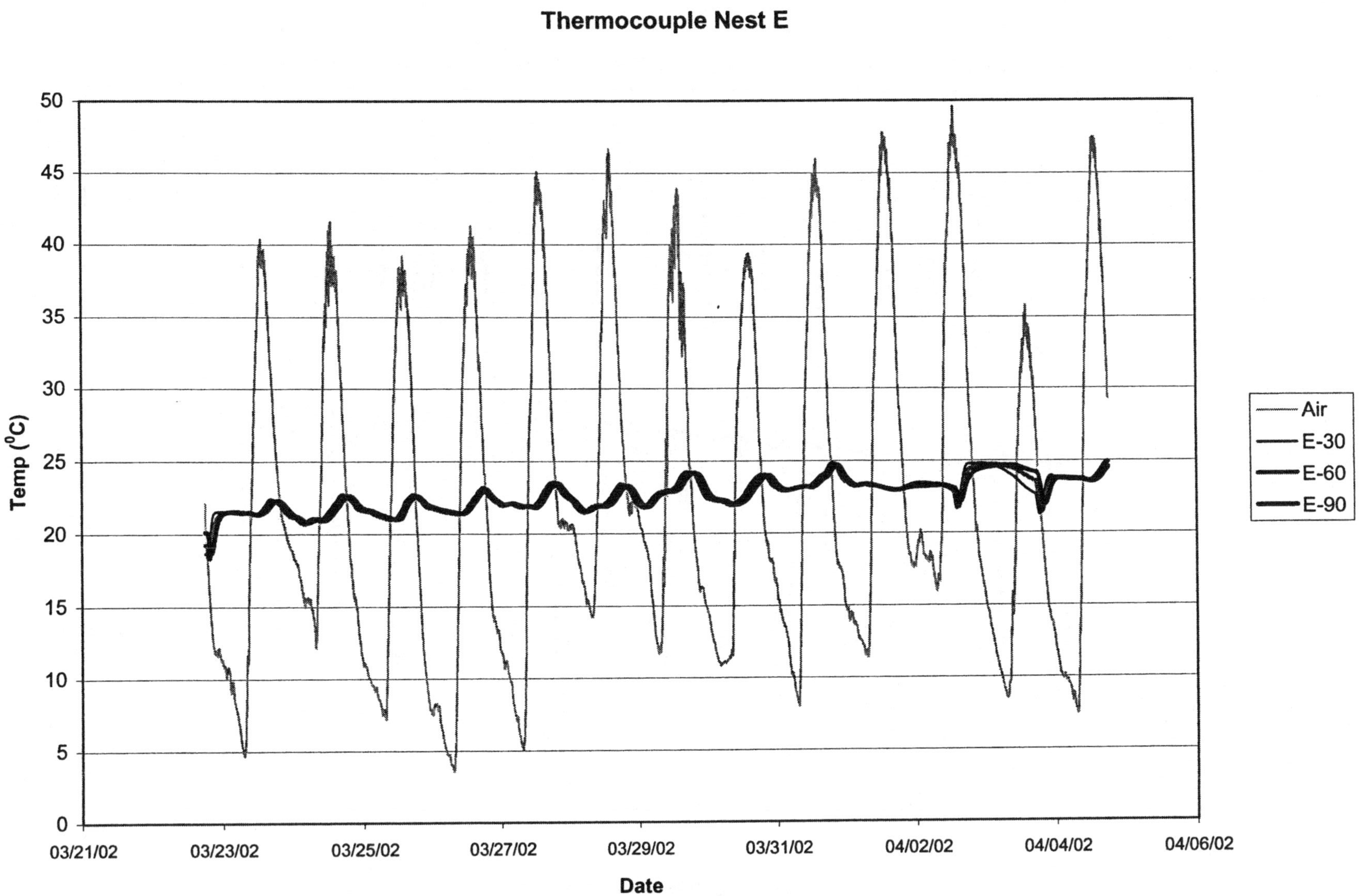

Figure G.5 (continued) Thermocouple data 3/23/02 through 5/15/02

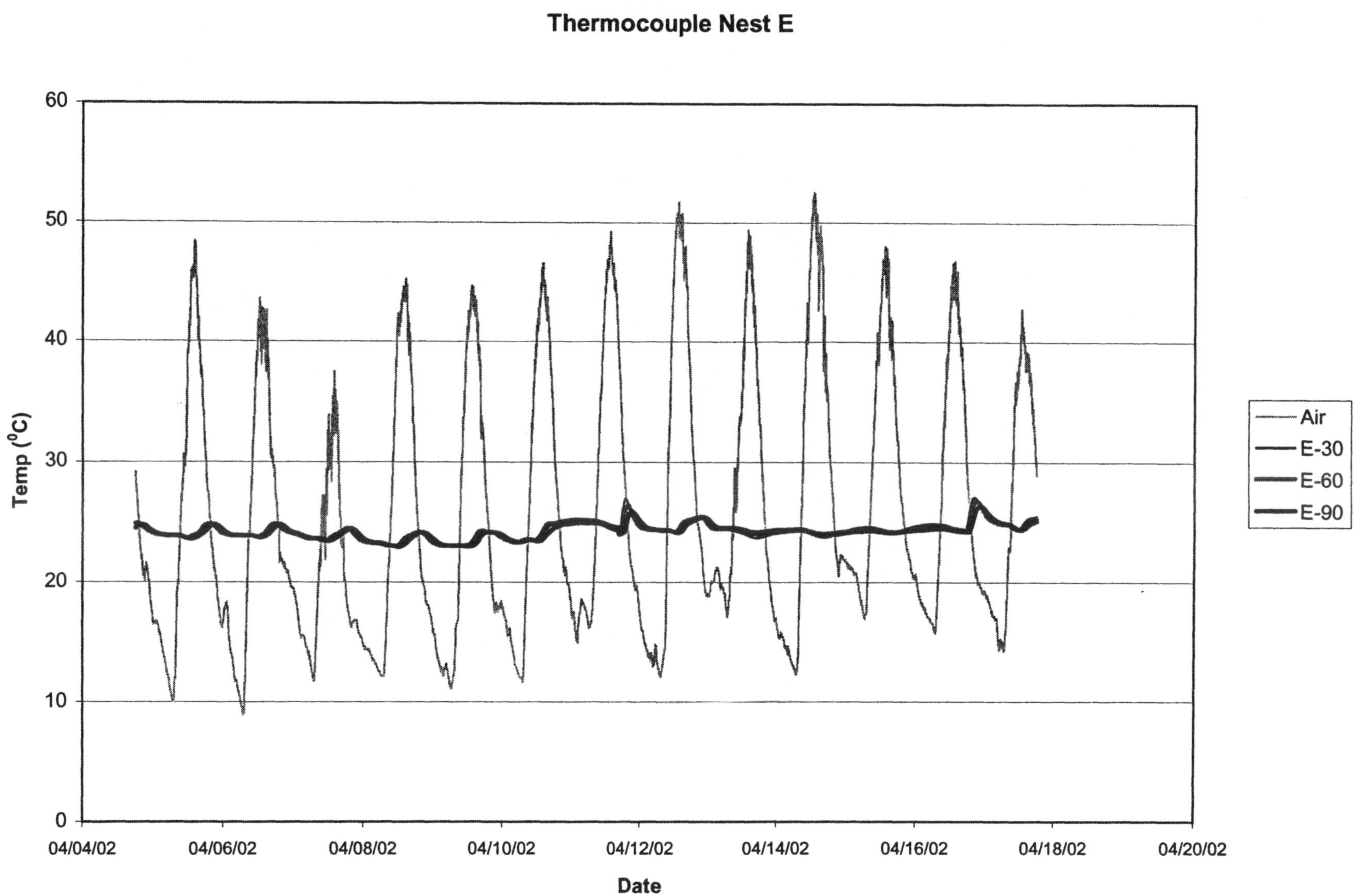

Figure G.5 (continued) Thermocouple data 3/23/02 through 5/15/02

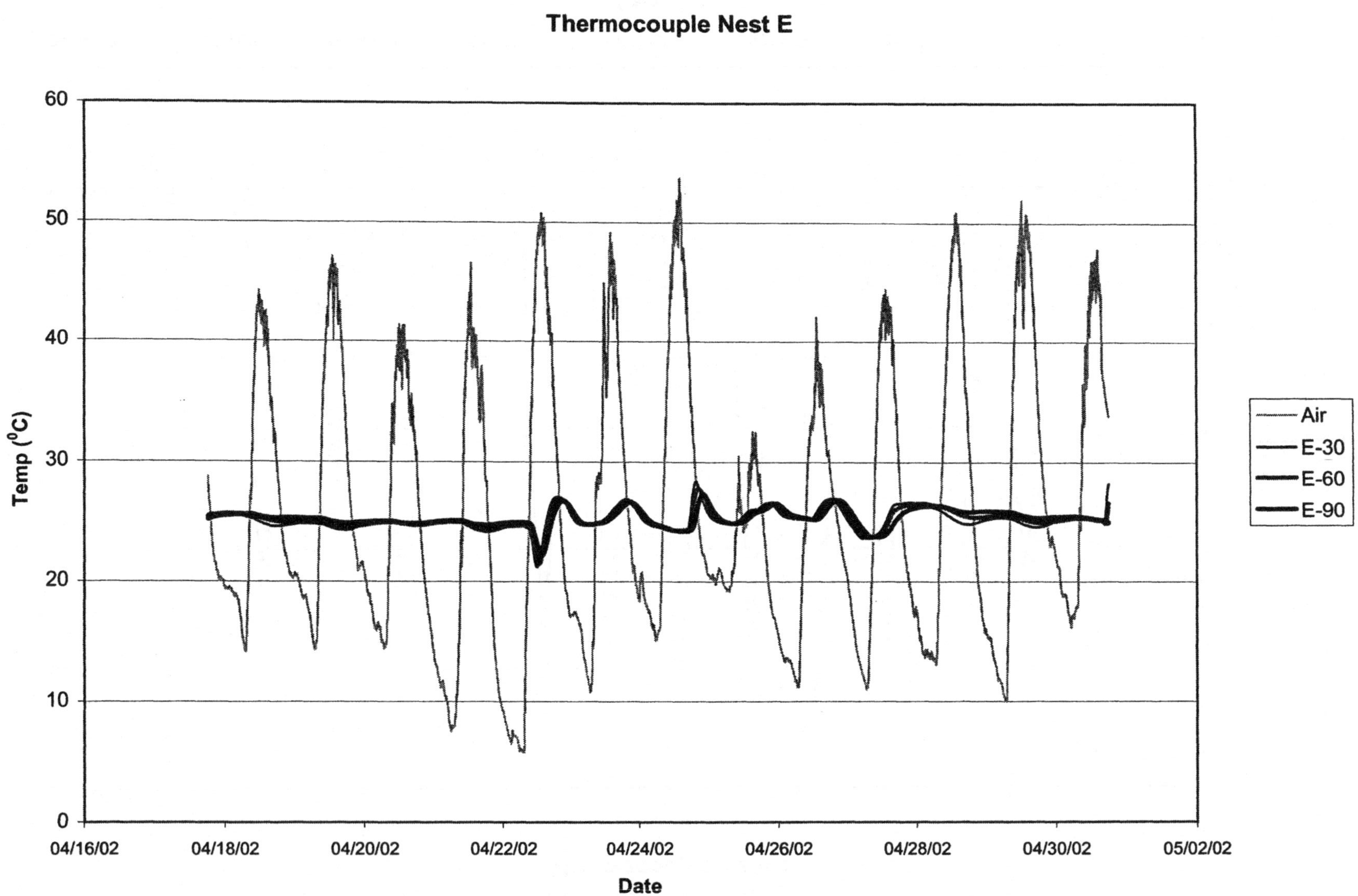

Figure G.5 (continued) Thermocouple data 3/23/02 through 5/15/02

Thermocouple Nest E

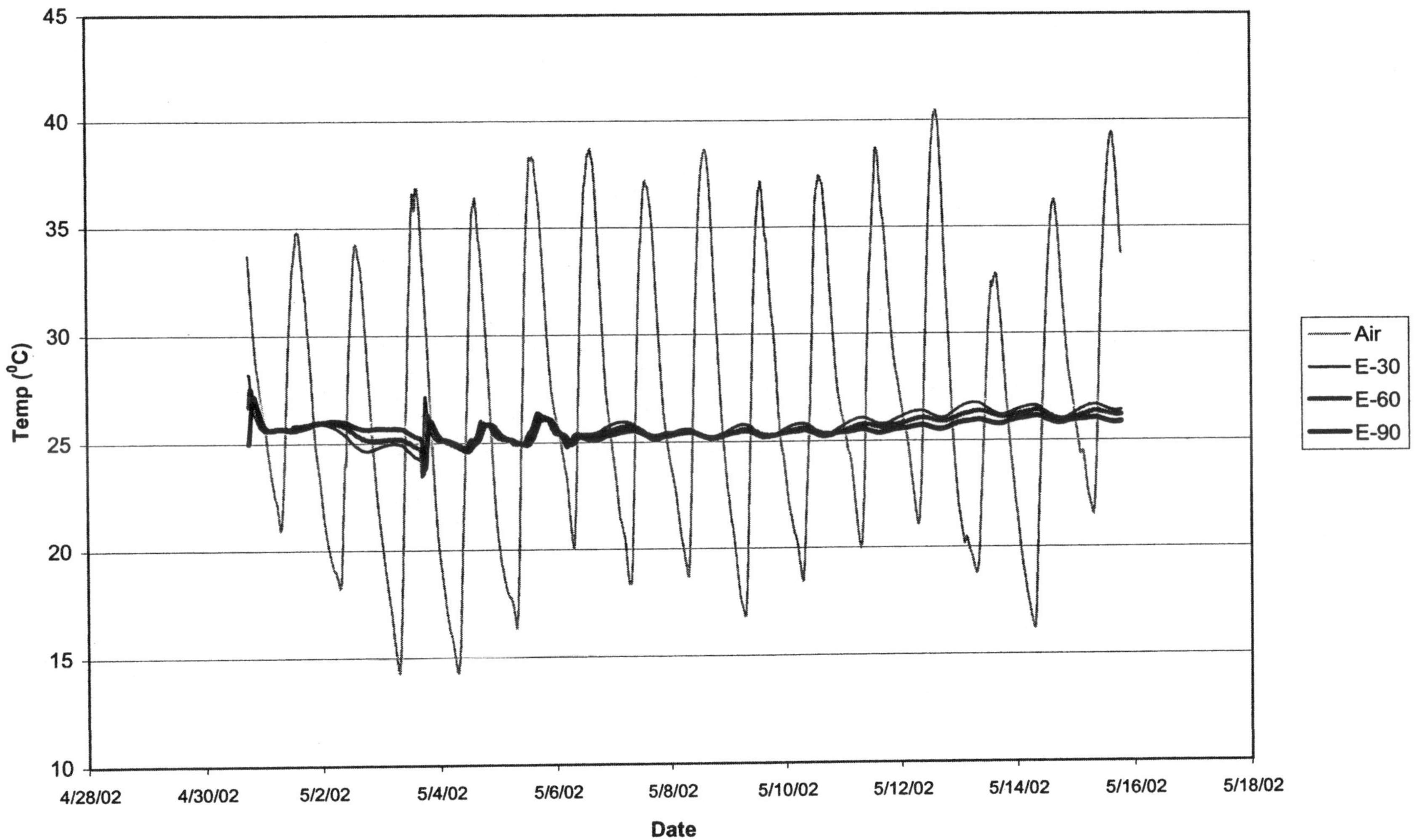

Figure G.5 (continued) Thermocouple data 3/23/02 through 5/15/02

APPENDIX H
CONVERSION TABLE

CONVERSION TABLE:
SI EQUIVALENT UNITS

	To Convert		
Description	**From Customary Units**	**To SI Units**	**Multiply By**
Area	acre	ha	4.047×10^{-1}
Flow	mgd	M^3/d	3.75×10^3
		or ML/d	3.785
Flow rate	gpm	m^3/s	6.309×10^{-5}
		or L/s	6.309×10^{-2}
Hydraulic conductivity	ft/day	m/day	3.048×10^{-1}
Infiltration rate	ft/day	m/day	3.048×10^{-1}
Length	in	mm	2.540×10
	ft	m	3.048×10^{-1}
Pressure	psi	kPa	6.895
Seepage velocity	ft/day	m/day	3.048×10^{-1}
Temperature	°F	°C	(°F – 32) × 5/9
Volume	gal	m^3	3.785×10^{-3}
		or L	3.785
		or ML	3.785×10^{-6}

REFERENCES

Ashworth, John B. 1990. Evaluation of Ground-Water Resources in El Paso County, Texas. Texas Water Development Board Report 324.

Boyle Engineering Corporation and Engineering-Science. 1994. Conjunctive Water Resource Management Technical Data Report, Volume I.

Boyle Engineering Corporation and Parsons Engineering Science, Inc. 1995. Study Report: Aquifer Storage and Recovery Investigation, Final Report, September 1995. Prepared for New Mexico-Texas Water Commission.

Boyle Engineering Corporation and Engineering-Science. 1997. *Aquifer Storage and Recovery: Supplemental Investigation of Recharge Basins*. New Mexico/Texas Water Commission.

Buszka, P., R. Brock, and R. Hooper. 1994. Hydrogeology and Selected Water-Quality Aspects of the Hueco Bolson Aquifer at the Hueco Bolson Recharge Project Area, El Paso, U.S. Geological Survey.

Constantz, J., R. Niswonger, and A. Stewart. 1999. The Use of Heat as a Tracer to Monitor Recharge Rates Beneath Streams and Recharge Ponds. Proceedings of 9th Biennial Symposium on the Artificial Recharge of Groundwater, June 10-12, 1999, p. 193-202, Tempe, Ariz.: Arizona Hydrological Society, Salt River Project, U.S. Water Conservation Laboratory, and Arizona Department of Water Resources.

Lapham, Wayne W. 1989. Use of Temperature Profiles Beneath Streams to Determine Rates of Vertical Ground-Water Flow and Vertical Hydraulic Conductivity. U.S. Geological Survey Water-Supply Paper 2337. Reston, Va.: Department of the Interior, U.S. Geological Survey.

Marsh, Floyd L., L. Dueker, M. ASCE, and G. Small. 1993. Recharge Well Technology at the Water Campus Project, Scottsdale, Arizona.

Marsh, Floyd L., G. Small, and C. Close. 1997. Vadose Zone Recharge Wells: Field Evaluation of an Innovative Design. Paper Presented at the 8th Biennial Symposium on the Artificial Recharge of Groundwater, Tempe, Ariz.

Marsh, Floyd L., G. Small, and C. Close. undated. Vadose Zone Recharge Well Testing: An Innovative Option for ASR.

Philip, J.R. 1992. Falling Head Ponded Infiltration, Water Resources Research, vol. 28, no. 8, p. 2147-2148. Washington, DC: American Geophysical Union.

Sperka, Roger. 1996. Unpublished East-West Cross-Section 3-3 through Well Site 41. El Paso Water Utilities, Public Service Board. 1996.

Stallman, R.W. 1965. Steady One-Dimensional Fluid Flow in a Semi-Infinite Porous Medium with Sinusoidal Surface Temperature. *Journal of Geophysical Research*, 70(12):2821-27. Washington, DC: American Geophysical Union.

Taniguchi, Makoto. 1993. Evaluation of Vertical Groundwater Fluxesand Thermal Properties of Aquifers Based on Transient Temperature-Depth Profiles. *Water Resources Research*, 29(7):2021-26. Washington, DC: American Geophysical Union.

Thomas, Carole L., A.E. Stewart, and J. Constantz. 2000. Determination of Infiltration and Percolation Rates Along a Reach of the Santa Fe River Near La Bajada, New Mexico. U.S. Geological Survey Water-Resources Investigations Report 00-4141. Denver, Colo.: U.S. Department of the Interior, U.S. Geological Survey.

U.S. EPA. 1999. The Class V Underground Injection Control Study, Volume 21, Aquifer Recharge and Aquifer Storage and Recovery Wells.

ABBREVIATIONS

ac	acre
ac-ft	acre-feet
AC	alternating current
ANSI	American National Standards Institute
ASR	Aquifer storage and recovery
AWWA	American Water Works Association
AwwaRF	Awwa Research Foundation
bgs	below ground surface
btoc	below top of casing
°C	degrees Celsius
cm	centimeter
EC	electrical conductivity
EPA	Environmental Protection Agency
EPWU	El Paso Water Utilities
°F	degrees Fahrenheit
ft	foot
ft/day	foot per day
FHWRP	Fred Hervey Water Reclamation Plant
gal	gallon
gpm	gallons per minute
ha	hectare
hr	hour

in	inch
K	Hydraulic Conductivity
kPa	kilopascals
L	liter
L/s	liter per second
m	meter
m^3	cubic meter
M	molar
M^3	cubic molar
M1	Monitor Well 1
M2	Monitor Well 2
M3	Monitor Well 3
M4	Monitor Well 4
M^3/d	cubic molar per day
m/day	meter per day
mg	milligram
mgd	million gallons per day
mg/L	milligram per liter
mL	milliliter
ML	megaliter
ML/d	megaliter per day
mm	millimeter
m^3/s	cubic meter per second
μS/cm	microsiemens per centimeter
O&M	operations and maintenance
P1	Piezometer 1

P2	Piezometer 2
P3	Piezometer 3
PAC	Project Advisory Committee
PRV	pressure reducing valve
psi	pounds per square inch
PVC	polyvinyl chloride
RTU	remote terminal unit
sec	second
TDS	total dissolved solids
UIC	Underground Injection Control
USDW	underground sources of drinking water
USGS	United States Geologic Survey

IWA Publishing's authorised EU representative for General Product Safety Regulations is Diane D'Arras, 15 rue Duret, 75116 Paris, France, e-mail: safety@iwap.co.uk.

Printed and bound by CPI Group (UK) Ltd, Croydon, CR0 4YY
17/09/2026
02218808-0005